KANT'S METAPHYSICS AND THEORY OF SCIENCE

Kant's Metaphysics and Theory of Science

by

GOTTFRIED MARTIN

TRANSLATED FROM THE GERMAN
BY
P. G. LUCAS

GREENWOOD PRESS, PUBLISHERS
WESTPORT, CONNECTICUT

Library of Congress Cataloging in Publication Data

Martin, Gottfried, 1901-
Kant's metaphysics and theory of science.

Reprint of the 1955 ed. published by the Manchester University Press, Manchester, Eng.
Translation of Immanuel Kant; Ontologie and Wissenschaftstheorie.
Includes bibliographical references.
1. Kant, Immanuel, 1724-1804--Ontology.
I. Title.
B2799.O5M313 1973 111 73-15053
ISBN 0-8371-7154-7

Translated from the German Edition
Immanuel Kant, Ontologie und Wissenschaftstheorie
[Cologne, 1951]

This edition originally published in 1955 by Manchester University Press, Manchester, England.

Reprinted with the permission of Manchester University Press.

Reprinted in 1974 by Greenwood Press,
A division of Congressional Information Service, Inc.
88 Post Road West, Westport, Connecticut 06881

Library of Congress catalog card number 73-15053
ISBN 0-8371-7154-7

Printed in the United States of America

10 9 8 7 6 5 4 3 2

PREFACE

The *Critique of Pure Reason* is a sea that is fed by two great streams: one stream is the new natural science, the other stream is the old ontology. It is the lasting merit of the Neo-Kantians to have shown the importance of the natural sciences and especially of physics for Kant. The *Critique of Pure Reason* is certainly a theory of physical science; it is certainly, as is shown at length in Cohen's presentation of it, a theory of experience. But the *Critique of Pure Reason* is not only a theory of mathematical natural science; it is at least to the same degree an ontology. That is why the old ontology could be taken up again in the *Critique of Pure Reason* and why, through this living renewal, new developments were possible. The aim of our interpretation is to show the inner connection between ontology and theory of science. Some will perhaps say that this is merely to add another to the many interpretations that exist already. But every new interpretation is only a new demonstration of the inexhaustibility of Kantian thought—the inexhaustibility that is the true sign of genius, that gives to Kant's works the same radiant splendour as it gives to the songs of Homer, the Bach Passions, the poetry of Goethe, and the same depth and breadth as it gives to the works of Plato, of Aristotle, of Thomas Aquinas.

Enquiries are presented in this work that have occupied the author over a long period of time. He recalls Gauss's saying that when a house is finished, the scaffolding that was needed to build it is taken down. The experts will easily see how much I am indebted to others. May the book be a new sign of the inexhaustible depth of Kantian thought.

G.M.

TRANSLATOR'S NOTE

Page-references are given in the text as follows :

Kant: for the *Critique of Pure Reason*, to the 1st and 2nd editions ('A' and 'B'); otherwise to the Berlin Academy edition. In the appropriate cases, the page-reference to T. K. Abbott, *Kant's Critique of Practical Reason and other works . . .* (' Abb.') has been added. A list of the sections of the *Critique of Pure Reason* which receive detailed comment will be found in the index, under ' Kant '.

Leibniz: to C. J. Gerhardt's editions of the philosophical works ('G') and the mathematical works ('GM').

Hegel: to Glockner's edition of the collected works.

The quotations from Aristotle follow the Oxford translation. In all other cases new translations have been made.

The English version has been discussed in detail with the author, and with his consent certain minor changes have been made.

CONTENTS

INTRODUCTION

THE PHILOSOPHY OF LEIBNIZ

AT the head of our enquiries we place Leibniz. In so doing we are following Kant's own procedure, which takes the form of a continuous discussion with Leibniz. The centre of Leibniz's philosophy is the doctrine of monads. A monad is a simple substance—so it is defined at the beginning of both the treatises on the monad. The term 'monad' takes up the Greek term *monas*. *Monas* means both unity as such and that which is single, primarily therefore substance in its unity (*Monadology*, § 1*f*., *Principles of Nature and of Grace*, § 1). This double meaning of *monas* is continuously effective throughout Leibniz's doctrine of monads.

The archetype of unity for the Greeks and for Leibniz is the living human being in the comprehensive unity of his existence. From this starting-point all monads are first defined as living. By a bold leap the whole universe is then filled with a dense sea of living things, that is to say with a dense sea of monads. Everything is alive; everything unfruitful, everything sterile, everything dead in the universe is only outward illusion (*Monadology*, § 69). To see into the things that seem to be dead is to become aware even there, in what seems to be dead, of an infinite fullness of living things. In this great ocean of living things there are no empty places. Wherever one looks there surges an infinite world of creatures, of living beings, of animals, of entelechies, of souls. Every single particle of matter, however small, is a garden full of plants, a pond full of fishes, and every twig of these plants in this garden and every drop in the blood of these fishes is another garden full of plants, another pond full of fishes, and so forth to infinity (§ 67). Everywhere, in the infinitely great and in the infinitely small, there is life, everywhere there are monads.

The two fundamental determinations of our being, namely thinking and willing, belong to every monad in their universal form as perception and appetition (§§ 15, 16). Every monad perceives and every monad wills, but not all of them in the same way, and thus there arises a graduated series of monads. The lowest grade consists of bare monads, whose spiritual faculties are completely obscure and confused. It is not very easy to say what Leibniz was thinking of, with these bare monads. Are they plants? Hegel thought they were inorganic monads (XIX,461). However that may be, plants occur as one of the lowest grades, then animals, then human beings, and on other stars there may perhaps be living beings still more finely organised than human beings here on this earth. All these are monads, living monads.

Is God a monad? In some places Leibniz seems to be hesitating to answer, but in general he answers yes. God is a monad *sui generis*, in distinction from all other monads, which are the created monads. Most of the people who have written about the doctrine of monads have answered the question in this way, for instance Hegel (XIX,464) among the older writers and Dillmann [1] among modern writers.

It then becomes a property solely of created monads to have a body and, not in its material identity but in its function, never to lose this body.

The monad becomes the central concept of Leibniz's ontology: for him all reality is involved in the existence of individual monads one by one. Real existence is only possible as the individual existence of a monad or of a modification of an individual monad. This is Leibniz's fundamental ontological thesis, and constitutes what has been called Leibniz's pluralism or individualism. It is most clearly expressed in the correspondence with Des Bosses, but it is present in the doctrine of monads from the outset, in the definition of a monad as a simple substance. From the earliest systematic investigations of the concept of substance onwards, the existence of substance has always been held to be real existence, properly speaking; at least whenever the concept of substance was taken broadly enough for the category of substance to be applicable to the existence of God. On this problem Leibniz may be said to have taken up the most radical standpoint that is possible at all. All reality is contained in the existence of the individual substance. Outside and between the individual existence of monads there is no reality.

Since only monads and modifications of monads have real existence in this sense, relations cannot have real existence but only an ideal existence, or to use an alternative expression which Leibniz often uses, they only have mental existence (G.II,486, to Des Bosses). One might say that in a certain sense they are only thoughts. They include relations in the narrower sense such as equality, similarity, difference, and causality; but they also include number, time, duration, space, the extension of bodies, all of which Leibniz counts as relations for the purposes of categorial analysis and of ontology, so that for him existence as a whole consists of substances, modifications, and relations (*New Essays*, II,1,3). Relations do not have real existence but only ideal existence; with this ontological thesis Leibniz takes up and develops doctrines which go back to antiquity. The doubts about reality that originate with the Sophists and the Sceptics apply particularly aptly to relations. The argument readily suggests itself that the existence of relations depends solely on their being represented. According to this conception, every human being has for example a determinate age and this property belongs to him at least in the first instance as a real property. But being older than or being younger than someone else belongs to him only when an observer compares the two men. Being older belongs to him only on the basis of this comparison, and being older is thus either itself only a representation, or at least it depends on a representation.

But this nominalist root, as one may call it, of the ideality of relation is not its only root for Leibniz, and perhaps not even its most important root. Besides these nominalist considerations, there is a group of problems worked out by the Scholastics in which relations are considered in connection with transcendental concepts. It had been shown that the conception of substances with accidents produces serious difficulties when the existence of relations comes to be defined. An accident, according to the definition of the concept, is always an accident of a determinate substance which carries its existence. Following this line of thought, Leibniz also holds that every modification is a modification of a determinate individual monad in whose substantial existence the existence of the modification is rooted. But when it is defined with more exact logic, a relation seems to lead to an accident that exists in two substances. Such a relation—to use a scholastic expression often borrowed by Leibniz, and used by him in the correspondence with Clarke (5th Letter, § 47)—would be in some ways an accident that had one foot in one substance and the other foot in another substance. Such a state of affairs is incompatible with the strict concept of an accident. Starting from this problem the Scholastics had developed the concept of transcendental relation, and finally Ockham had defined all relations as transcendental relations. A transcendental relation has no reality—no distinct reality in the sense and only in the sense in which the *unum transcendens* also has no distinct reality.[2] Leibniz denies the reality of relations, and this denial contains not merely an element of scepticism but also an element of transcendental philosophy and thus of Aristotelianism; relations have reality in a certain sense after all, not the reality of substance but the reality of transcendentals. This real character of relations is strengthened by a further element that could perhaps be called Platonic. It is true that relations have only an ideal, a mental existence; their being consists in their being thought by an understanding. But the understanding that thinks relations is the understanding of God. Relations have their substantial reality taken away from them by being referred to an understanding, but because the understanding that carries them is the divine understanding, they can be said in a certain sense to receive back again a new reality. This argument will become clearer when we consider concrete particular problems. With a view to what we have to say later, we choose matter and space.

Matter, as Leibniz keeps repeating, is a mere phenomenon, and it is even clearer for him that space is a mere phenomenon. Investigation of geometry by categorial analysis shows space itself to be purely a problem of relations; all spatial properties are relations, as is clearly shown when the concepts and judgments of geometry are analysed. Space, being a system of relations, offers all the possible spatial relations between bodies, hence between monads. The spatial relations between monads, in which the problem of space issues, are not themselves monads, nor can they be modifications of monads, since every modification can only be a modification of a particular monad and no modification can subsist between two monads. A spatial relation, for example that A lies between B and C, is a relation between several monads and hence has no real existence. This being-between consists only in its being represented, but in its being represented by the divine understanding. Space therefore becomes a phenomenon, but a *phaenomenon Dei* (G.II,438). This phenomenal character, with its fundamental reference to the divine understanding, holds of all relations, of relations in the narrower sense as well as of number, space, extension, bodies, time, duration (G.II,435).

This distinction between the phenomenal existence of relative properties and the real existence of monads and their modifications is the foundation for the distinction between the realm of nature and realm of grace. The distinction has its origin, especially in the

Monadology, in the idea of a divine republic which embraces all thinking spirits. This realm, the realm of spirits, forms the most perfect state under God as the most perfect ruler. It also forms a moral world within the natural world. The physical realm of nature and the moral realm of grace are contrasted realms. The physical realm of nature is the universe regarded as a machine, the realm of grace is the divine republic of spirits. The pre-established harmony connects the two realms, so that all events run parallel. Through the pre-established harmony the course of the universe as determined by natural laws—the universe regarded as a machine—is kept in exact agreement with the course of events in the realm of grace, the divine republic of spirits (*Monadology*, § 87).

These two realms may be identified in the main with the real realm of monads and the phenomenal world of space, time, motion and bodies. By means of this metaphysical distinction between the real realm of monads and the phenomenal realm of bodies in space and time, Leibniz resolves the conflict between the old demands of theology and ethics and the new demands of natural science. The old problems demand freedom of moral and religious decision, the new results of natural science demand thoroughgoing causal determination of the machine of the world—*de la machine de l'univers* (§ 87). The ontological distinction between the real realm of monads and the mental realm of bodies in space and time contains the direct solution to the conflict. In the real realm of monads there is freedom, while natural causality rules the mental realm of bodies in space and time. A separate region is thus created for each of the two great systems of laws, and finally the pre-established harmony sees to it that no contradiction can ever arise between the two regions and that complete harmony reigns between them always.

The foundation of this system is the presupposition that the world is created by God, and this presupposition combines the problems of natural theology with those of ontology and of the foundations of mathematics and physics. The creation of the world by God is interpreted to mean that the infinity of possible worlds is grounded in God's thinking and the best of all possible worlds is called into existence from the infinity of possible worlds through God's will.

The presupposition that the infinity of possible worlds has its being in being thought by God is the foundation on the one hand for a proof of the existence of God, and on the other hand for the possibility of mathematics and physics. God exists and his existence is provable because the infinity of possible worlds could only be thought by an infinite understanding. This is how Leibniz meets the old demand of natural theology for a proof of the existence of God. But the divine thinking also provides the ground of the possibility of mathematics, because in his continuous thinking of possible worlds God also thinks continuously the eternal truths that hold for these worlds, in particular all numerical and spatial relations. This holds especially of Euclidean space which, as the only possible space, is the same for all possible worlds. The being of Euclidean space thus consists primarily in its being thought without contradiction by God.

The presupposition that God thinks all possible worlds is also the ground for the possibility of physics. If all possible worlds and among them this world, are continuously thought by God, the being of the world is primarily a being thought and the world is therefore in its original being intelligible, transparent to reason. It may still be the case that real insight into the world is only possible for an infinite understanding, and that human thinking is finite and limited; but all this does not prevent the existence of the world from being in principle rational and hence conceivable and understandable by human beings, at least through an infinitely extended approximation.

As well as assuring the possibility of mathematics and of physics from the side of the known, the same presupposition also assures it from the side of knowing. Knowing is primarily God's knowing and all human knowing is thus only a repetition of the primary divine knowing. Thus, to return to a concrete problem, the propositions of three-dimensional Euclidean geometry are primarily and continuously thought by God, and when the mathematician discovers, understands, and proves an adequately formulated proposition in geometry, this knowing is a repetition of the primary divine knowing. Leibniz's universality, the fundamental feature of his character, shows itself in the fact that he takes up the leading ideas from all the great philosophical systems. The profound consequences of Leibniz's connections with Plato, which we first met when we were considering the problem of the existence of relations, begin to unfold at this point. Plato pronounced repeatedly that thinking the truth means becoming like God. Leibniz may have met the ancient Platonic thesis in its most palpable form in Malebranche, in the proposition that we see all things in God. If Leibniz is unable to suppress some criticism of this far-reaching proposition of Malebranche, he agrees with it essentially, as he is bound to do if he holds to his presupposition that God thinks all possible worlds.

We certainly do not wish to ignore that when he referred scientific knowledge to the divine understanding Leibniz was carrying on a great tradition, following Plato, Plotinus, Augustin, and Malebranche. And it is certainly understandable that at a time of very great scientific discoveries—one need only think of the infinitesimal calculus and classical physics—those who theorise about science should incline to a splendid conception of scientific knowledge, and it may be that this splendid conception of scientific knowledge as an approximation to divine thought gave wings to science to make new discoveries. But the old question will always recur, whether this splendid definition of knowing is not all too splendid, whether the optimism that shows itself here is not all too optimistic.

This may also hold of a further theologico-ontological problem which Leibniz believed he had solved, the problem of the theodicy, of justifying God's ways to man. How does it stand with the evil in the world? Or as a purely theological question, how can God permit evil? The solution to this question also follows for Leibniz from his presupposition and its consequence, that the actual world is the best of all possible worlds. If this world was selected from among an infinity of possible worlds it can only be the best of all possible worlds, since God can only choose the best. The optimism which we found in his definition of knowledge is intensified here to a degree which is almost beyond understanding. Criticism, indeed scorn, of Leibniz began at once over this point and Voltaire's comic novel *Candide* is the wittiest of all satires on a philosophical idea.

But with all this we must not fail to see that here we have an attempt at a comprehensive synthesis which has the quality of genius. Leibniz is not only in himself the last universal scholar. The comprehensive unity of the sciences is his deliberate aim, and Leibniz strives for unity in the first instance as a synthesis of theology, metaphysics, mathematics, and physics. Hegel, in tones of unmistakable scorn, called this synthesis a metaphysical novel (XIX,454), though we cannot help asking what would be the corresponding title that Hegel's own thought-construction deserves.

The real discussion of Leibniz's philosophy is, if we are right, the philosophy of Kant. Kant poses in all seriousness the question, whether these definitions of being, of knowing, and of evil, are not all too optimistic. The Kantian philosophy is in intention and in achievement a fundamental discussion with Leibniz, a discussion in which agreement and opposition are indissolubly connected. Kant can therefore justly say, at the end of the polemic against Eberhard (VIII,250), looking back on his philosophical life-work, 'The *Critique of Pure Reason* is, after all, the real apologia for Leibniz, even against his followers who exalt him with praises which do him no honour.'

PART ONE

UNITY

CHAPTER I

SPACE AND TIME

§ 1

The Problem of what Space is

THE Transcendental Aesthetic begins with the question:

> What then, are space and time? Are they real beings? Or are they only determinations or relations of things, but yet such as would belong to things in themselves even if they were not intuited, or again such as only attach to the form of intuition and hence to the subjective nature of our mind, apart from which these predicates could not be ascribed to anything whatsoever? (A 23, B 37).

The question that Kant is asking is not how we can know space and time, but quite simply what space and time are. This question has an ontological sense, and there is a close connection between the way it is asked and the methods of ontological enquiry that had been worked out in the course of history; this can be seen even more clearly in the *Inaugural Dissertation* of 1770, *De mundi sensibilis atque intelligibilis forma et principiis.* Kant says there first of time, and then in almost exactly the same words of space, 'Space is not something objective and real, neither substance nor accident nor relation, but something subjective and ideal issuing by a constant law from the nature of the mind' (II,400,403). There are, he is saying, four possibilities. Space is either a real being, a substance; or it is a determination, an accident; or it is a relation; and it can be a relation in two ways, either as a relation subsisting in itself, an objective relation, or as a relation that originates in our mind and attaches to the subjective nature of our mind, a subjective relation therefore, if we may go a little beyond the terminology of the *Dissertation.*

Kant is asking about the being of space and time, and the framework for his answer is provided by the four possible kinds of being developed in ontology. The whole of being, according to ontological doctrine as Kant understands it, consists of these four possibilities, and space and time must therefore be included under them. It is true that the connection between Kant's doctrine and the old ontology is not apparent at first sight. Kant regards it as already firmly established that as regards space the three first possibilities, namely substance, accident, and real relation, have been finally disposed of, and that it was Leibniz who disposed of them. Kant does not therefore think it necessary to discuss the first three possibilities; but all the same one occasionally finds references to them, for instance when he says,

> Those who maintain the absolute reality of space as subsisting . . . have to admit two eternal and infinite non-entities (space and time) subsisting in themselves, which are there (yet without anything real being there) only in order to contain in themselves all that is real (A 39, B 56).

Two eternal entities subsisting in themselves with substantial and absolute reality are in fact substances, except that on consideration one sees at once that such substances would not be entities but non-entities.

It can be shown from Kant's manuscript material that when he alludes to the ontological view of space and time as substances, he is thinking primarily of Epicurus.[1] It is in fact primarily in atomistic doctrines that space and time are made out to be substances. In a sense, it could even be said that from the atomistic standpoint which is taken up in Plato's *Timaeus*, space is regarded as a substance. If the definition of space given in the *Timaeus* (49a) is considered from the point of view of the Aristotelian conceptual scheme, it cannot be conceived in any other way. In the atomistic doctrines of Democritus and of Leucippus, as can be well understood from consideration of the problems involved, space is actually defined as a substance. Atoms and empty space are the fundamental concepts of the atomists. The problem is best seen in Aristotle's discussion of space in the fourth book of the *Physics*, though the doctrine probably had its greatest historical influence in the versions of Lucretius and Epicurus. The struggle with the doctrine that space is to be defined as a substance continues throughout the whole history of philosophy. For Kant the correspondence between Leibniz and Clarke, which was in substance a correspondence between Leibniz and Newton, was probably particularly important. From the atomistic standpoint space and time are regarded as substances, that is to say as two eternal and immutable things which contain within themselves atoms and their configurations, and thus everything real. These two substances would still exist if all real things were taken out of them, and we have seen how Kant alludes to this (A 39, B 56). It is a point of view that becomes especially easy to grasp when it is combined with a cosmogony.

Space and time would then always have existed, and at the creation things were put into time and space, both of which had always been there. This holds particularly for Newton.

But weighty objections can be sustained against a definition of space and time as substances. The decisive consideration was found by Zeno. Zeno said, according to Aristotle (*Physics* IV.1.209a23*f*.), that if everything real is in space, then obviously space, if it is real, must also be in space, and there would therefore have to be a space of spaces and so forth to infinity. The argument is that everything real is in space; anything that is at any point in time and in any position in space, that at all events is real. That the sphinx, the stock Greek example of something unreal, is not real is shown by the fact that it does not exist anywhere in space and time. This is certainly the standpoint that the atomists have to take; atoms and their configurations, to which everything real can be reduced, are in space, and thus the atomists are heavily committed to the doctrine that everything real is in space. But then, as Zeno's paradox shows, space itself can only be real if it is itself in space in its turn, and this is obviously nonsense.

An important ontological point was brought out for the first time in this paradox of Zeno. Space, as the paradox shows, must have a different being from that of things that are in space. Space cannot be anything like the spatial things which are in space; space itself cannot therefore be a spatial thing. Zeno's paradox shows for the first time that the question as to the being of space contains a special problem for ontology; space is at all events not itself a spatial thing, and in particular space is not a substance.

In the correspondence with Clarke, Leibniz places special emphasis on this point. It then emerges in the discussion that a second argument can be added to this ontological argument of Zeno on Christian grounds. If space and time are regarded as substances, both are eternal and infinite. But then there would be besides God, whose being is primarily that of an eternal and infinite substance, two further eternal and infinite substances, and under the Christian presuppositions of the correspondence this is impossible (4th Letter, § 10). Not only on the general ontological grounds of Zeno's paradox, but also on grounds of Christian dogma, space and time cannot be regarded as substances.

For Newton the dogmatic argument has a special significance. On physical grounds Newton requires the absolute reality of space. But if it is impossible to define space as a substance, the only alternative—if its absolute reality is to be retained—is to say that space is an accident. Since space and time are eternal and infinite, they can only be accidents of a substance which is itself eternal and infinite. This means that they can only be accidents of God. Newton cannot escape this consequence, and we believe that when Clarke defined space and time as qualities of God (4th Reply, §10) [2] he was correctly formulating Newton's point of view. Kant too is thinking of Newton when he considers the view that space and time are accidents.[3] But this definition of space and time as qualities of God is obviously dubious and a pure evasion, and Leibniz loses no opportunity to refer again and again to these perplexities and embarrassments (3rd Letter, § 3; 4th Letter, § 10). The ontological characterisation of space and time as properties or accidents has also, therefore, to be given up.

Of the possibilities that Kant had mentioned when he asked his question, the first two—space and time as substances (real beings) or as accidents (determinations)—are, as we have seen, untenable. Space must therefore be a relation. Aristotle had already brought out the relational character of space and time, and when Ockham fully worked out the problem he reached the same position.[4] The final definition of space as a relation was completed by Leibniz. There were sufficient grounds in categorial analysis to make him subsume space under the general concept of relation. Geometry, like the whole of mathematics, is a doctrine of relations; mathematics, as Leibniz saw quite clearly, is theory of relations (G. II,304, to Des Bosses). In this analysis the ontological definition of space is already foreshadowed. But the being of a relation still has to be defined, and here Leibniz had before him the scholastic distinction between *relatio realis*, *relatio transcendentalis*, and *relatio rationis*. This distinction was prepared by Thomas Aquinas, was developed expressly by Duns Scotus and Ockham, and was given a particularly precise form by Suarez. Leibniz probably found it in Suarez. A real relation is a relation which is distinguished from its fundaments by a real distinction (*distinctio realis*), that is to say, that is distinguished from its fundaments in the same way as one real thing (*res*) from another thing.[5] A *relatio rationis* is a relation which exists solely in representation. A transcendental relation is a relation which is not distinguished from its fundaments by a real distinction but which also does not exist solely in representation. Transcendental relation is further defined in different ways by Thomas Aquinas, Duns Scotus, and Ockham.[4] It can at all events be taken in one way as real relation, in another way as *relatio rationis*.

Leibniz denies that there is such a thing as real relation. A relation connects two fundaments, preferably and in any case as its final ground, two substances. A real relation would therefore be something real that had one foot in one substance and the other foot in another substance. It would be impossible for a relation to exist in this intermediate way between two real substances and relation cannot therefore be a separate real thing (G.II,517, to Des Bosses). In the same correspondence Leibniz finds the happy phrase [*tempus et spatium*] *sunt ordines, non res* (G.II,515). Real existence, in the sense of the existence of a particular real thing, is identical for Leibniz with the existence of monads or of their individual modifications; only monads are real in the specific sense that every monad is a real thing in itself (*res per se*). Space and time are therefore not themselves things, not themselves *res*, because otherwise they would be monads. They are ordered connections of monads or systems of orders of monads. As such they are not things but representations, and do not have real existence but ideal existence (G.II,486). Although Leibniz taken literally seems to define space as a *relatio rationis*, in substance he is developing the idea of transcendental relation.

This return to an objective existence for space and time is expressed by saying that the thinking for which space and time, taken as relations, are representations is God's thinking. Leibniz's theory of space reaches its climax in his ontological definition of space and time as *phaenomena Dei* (G.II,438). Kant in his pre-critical writings took the phenomenality of space and time in the same way, by referring them to God's thinking.

Let us review these doctrines of Leibniz once more, so that we can have them clearly before us. For Leibniz the only real being is that of monads and their individual modifications. Relations between monads are not themselves real, because otherwise they would have to be monads themselves. They only have ideal being, and they have their ideal being as representations in God's reason. God's thinking perceives all relations between monads continuously, and it is in being perceived in God's thinking that relations have their being. But although space and time as systems of relations only have ideal being, this gains a rich significance from its reference to the divine *ratio*, and the phenomenality of space and time, interpreted in this way, once more comes near to the being of ideas in so far as they are regarded by Plato and Augustin as God's thoughts.

§ 2

The Axiomatic Character of Geometry

Our guiding principle is that the *Critique of Pure Reason* is fed by two great streams: by theory of science and by ontology. Kant's doctrine of space and time makes this particularly clear. The Transcendental Aesthetic is certainly a discussion with Leibniz and Newton, as has often been shown. Not enough attention has however been paid to the fact that Leibniz and Newton are not only founders of the new natural science, but that they also took up the old philosophy, carried it further and developed it. Of the two, it seems to me that Leibniz is more important than Newton for Kant's thought, presumably because the philosophical problems are nearer to the foreground in Leibniz. At all events, if we are to understand Kant's exposition, method, and solution, his enquiries into the problem of space and time have to be approached from Leibniz's philosophy.

This is not to say that Kant did not oppose Leibniz over points of great importance and that Kant did not reach his position in certain ways precisely by struggling against Leibniz. This is certainly the case in respect of the question whether geometry is axiomatic in character. The problem is an old one but the foundations for a solution were first laid down by David Hilbert. Euclid's *Elements*, the earliest surviving textbook of mathematics and continuously used as a textbook from the first, have an axiomatic structure. First there are the fundamental definitions, then there is a series of propositions called axioms and postulates, and from these definitions, axioms, and postulates, Euclid demonstrates the theorems of geometry, arithmetic, and theory of proportion. Understandably, the question at once arises for mathematicians, logicians, and philosophers as to what this axiomatic structure means. The axioms are propositions which have the same form as the theorems that are deduced from them. But what kind of propositions are they? Are they propositions whose demonstrations have not yet been found, so that their validity only has to be claimed axiomatically as a temporary aid, until the day when the proof is found? Or are they genuine axioms, propositions that can be never proved, propositions which are perhaps self-evident? The discussion has gone on for two thousand years and both the possible points of view have been energetically represented.

Leibniz denies the axiomatic point of view. He is convinced that all mathematical propositions can be proved. This means that all mathematical propositions can be demonstrated from definitions and the principle of contradiction (2nd Letter to Clarke, § 1). Leibniz gave himself a lot of trouble, as can be seen from many passages (e.g. G.M.V, 183*f.*), trying to find proofs for the fundamental propositions which are usually assumed as axioms—as we now know, in vain.

The discussion itself turns in the first instance on the axiom of parallels, because this axiom is not used until Proposition 29 of the first Book of the *Elements*. The first twenty-eight propositions can be proved without using it, and there was therefore a case for asking about the demonstrability of this axiom first. The fundamental principles of the problem were seen comparatively early. There is a series of propositions which are equivalent to the axiom of parallels, so that one of these equivalent propositions may be proved instead. The proof itself would have to be an indirect proof, by taking the opposite of the axiom of parallels or of an equivalent proposition and then showing that there is a contradiction either in the opposite proposition or in the consequences that can be drawn from it. People looked for a long time; a comprehensive attempt was made round about 1730 by the Italian mathematician, G. Saccheri, and published in his book *Euclides ob omni naevo vindicatus* (Milan, 1733). Saccheri started from one of the equivalent propositions; he assumed that the sum of the angles of a rectangle is less than four right-angles. To his great astonishment Saccheri saw that he could develop a long chain of consequences from this 'false' presupposition, and he only found the contradiction that he was looking for after a very long time. When his investigation was checked it was very soon found that the contradiction was the result of a mistaken inference, and that the consequences of the 'false' presupposition were as far as could be seen free from contradiction. A system of theorems can be built on the presupposition that the sum of the angles of a triangle is less than two right-angles, just as the Euclidean system can be built on the presupposition that the sum of the angles of a triangle is equal to two right-angles. With this discovery the first non-Euclidean geometry had been found. One of the first to practise it was Lambert, the Berlin mathematician who was a friend of Kant.[1]

In the discussion about the axiomatic or non-axiomatic character of geometry Kant takes up the axiomatic standpoint decisively. How Kant reached this conviction has not yet been explained in detail. Perhaps Lambert himself was the link, and the problem of four-dimensional space is probably also important. Leibniz regards all the properties of Euclidean space as necessary in thought, including the three dimensions. In the *Theodicy* (§ 351) among other places there is a 'proof' of the three-dimensionality of space. Kant recognised the incorrectness of this proof in the work that he wrote when he was barely twenty called *Thoughts on the True Estimation of Living Forces.* In this his first work there is a place (frequently noticed) where he demands as the true geometry a geometry of any number of dimensions (I,23*f.*). In the critical writings the axioms of geometry are repeatedly called genuine axioms (A 732, B 760). The discovery of non-Euclidean geometries called up a storm of indignation among both mathematicians and philosophers, and nineteenth-century Kantians (and many even later) took part in the storm lustily. But it has been finally recognised that from the Kantian standpoint this was wrong, and Nelson, Meinecke, and Natorp have shown conclusively that under the Kantian presuppositions it is not only possible but necessary to assume the existence of non-Euclidean geometries.[2] There can be no doubt of the correctness of this view although closer investigation of particular points is desirable. I was myself able to show in my dissertation that Kant also recognised the axiomatic character of arithmetic, indeed, that Kant discovered the axioms of arithmetic through original mathematical enquiries, and that the axiomatic theory of arithmetic starts from these Kantian discoveries.[3]

Kant's way of formulating the view that geometry is axiomatic in character is to say that geometrical judgments are synthetic.[4] Analytic judgments and, correspondingly, synthetic judgments are defined by Kant in two ways. An analytic judgment is, first, one in which the concept of the predicate is contained in the concept of the subject, and secondly one that can be proved from the principle of contradiction alone. A synthetic judgment, correspondingly, is one in which the concept of the predicate is not contained in the concept of the subject, and one that cannot be proved from the principle of contradiction alone. It has often been complained that this Kantian distinction between analytic and synthetic judgments is very difficult to support with examples and that examples of analytic judgments are especially hard to find. Analytic judgments are in fact reduced for Kant almost exclusively to tautologies. The only analytic judgments that come to hand are certain general principles of the theory of quantities, e.g. the proposition that the whole is greater than its part.[5] The Kantian distinction between analytic and synthetic judgments can only be made clear by drawing on Leibniz's theory of mathematical judgments for examples of analytic judgments; and this is in fact the historical connection. Leibniz regards all mathematical judgments as analytic, to use Kantian terminology. They can be proved, according to Leibniz, from the principle of contradiction (taken together with definitions), and in geometrical judgments, as in all other judgments, the concept of the predicate is contained in the concept of the subject. The two Kantian definitions of analytic judgments are both involved in Leibniz's theory of judgments. That is why the proposition that the sum of the angles of a triangle is two right-angles can be proved for Leibniz from the principle of contradiction alone, and why the concept of the predicate 'sum of the angles equal to two right-angles' is contained for him in the concept of the subject 'triangle'.

If the axiomatic conception of geometry is adopted, the problem is different. In this case, the proposition that the sum of the angles of a triangle is two right-angles cannot be proved from the principle of contradiction alone. The concept 'sum of the angles equal to two right-angles' is not contained in the concept 'triangle' and is not in a necessary logical connection with it, since the two other logically possible predicates 'sum of the angles less than two right-angles' and 'sum of the angles greater than two right-angles' can also be thought as joined to the subject 'triangle'. Thus the subject 'triangle' is connected in Euclidean geometry with the predicate 'sum of the angles equal to two right-angles', in Bolyai-Lobatschevskian geometry with the predicate '. . . less than two right-angles' and in Riemannian geometry with the predicate '. . . greater than two right-angles'. We thus interpret axiomatic theory as confirming the Kantian thesis as to the synthetic character of geometrical judgments, at least in so far as Kant bases the synthetic character of geometrical judgments on the axiomatic view, that is to say in so far as Kant starts from the twofold conviction that geometrical judgments cannot be proved from the principle of contradiction alone, and that the connection between subject and predicate in the judgments of Euclidean geometry is not logically necessary, the subjects of these judgments being capable of being connected with other predicates.

By calling geometry synthetic in this first sense, namely the axiomatic sense, the spontaneity of geometrical thought is also brought to light for the first time. The axioms can be conceived as insights into the essential nature of space, but they can also be conceived as free postulates of thought. Nearly all the evidence goes against the conception of axioms as insights into the essence of space, especially the high degree of freedom with which axioms can be postulated and altered. Modern axiomatic work scarcely permits any other way of conceiving axioms than as free postulates. Hilbert's axiomatics begins with the pronouncement, 'We think three different systems of things.'[6] In the axioms, freely postulated qualities are then attributed to these freely thought things. Hence the concepts 'sum of the angles equal to two right-angles', '. . . less than two right-angles', and '. . . greater than two right-angles' can all be attributed to the concept 'triangle'. Spontaneity of thought is apparent in this free choice of approach, and spontaneity of thought shows itself again in the further development. When a system of axioms has once been constructed, endless new systems of axioms can always be constructed out of it by omitting particular axioms or adding them (if the system is not already complete), or by changing particular axioms and then making new omissions and additions. There is thus an unlimited number of geometries, of arithmetics, of logics, and Hilbert's construction, rich as it is, represents in spite of its overwhelming wealth only a small section of an undertaking that can be continued indefinitely. This phenomenon can only be understood if thought is regarded as productive, and the attempted interpretations that start from modern axiomatic work have mostly taken it this way. The only question is whether these interpretations have not perhaps gone too far in maintaining that the freedom of axiomatic thought has to be understood as mere play or as mere convention.

The connection between Kant's work and axiomatic work is very important for us. According to our interpretation, one meaning of the Kantian thesis that geometrical thought is synthetic and productive in character, and at the same time one ground for it, can be found in the axiomatic character of geometry.

§ 3

The Constructive Character of Geometry

The constructive character of geometry provides Kant with a second reason for calling geometrical judgments synthetic. The constructive character of geometry and of mathematics in general is according to Kant a fundamental feature of mathematics which distinguishes it from philosophy. Kant develops this distinction. He gives as a definition, 'Philosophical knowledge is knowledge by reason out of concepts, mathematical knowledge is knowledge by reason out of the construction of concepts. To construct a concept means to exhibit the a priori intuition that corresponds to it' (A 713, B 741).

The question now is what this constructing in intuition means, and on this subject Kant has almost always been misunderstood. This constructing certainly cannot mean drawing on the blackboard. Kant himself mentions in the polemic against Eberhard that the circle that is drawn on the board or in the sand is quite certainly not a circle (VIII,191*n*.). And yet however inaccurate the circle in the sand may be, it stands for a circle for the mathematician, and the mathematician can demonstrate all the properties of the circle from it precisely. The meaning of construction becomes apparent if we consider proofs and definitions.

When the mathematician wants to prove a proposition he always does so by means of constructions.

> [The geometer] at once begins by constructing a triangle. Since he knows that the sum of two right-angles is exactly equal to the sum of all the adjacent angles which can be constructed from a single point on a straight line, he prolongs one side of his triangle and obtains two adjacent angles, which together are equal to two right-angles. He then divides the external angle by drawing a line parallel to the opposite side of the triangle, and observes that he has thus obtained an external adjacent angle which is equal to an internal angle—and so on. In this fashion, through a chain of inferences guided throughout by intuition, he arrives at a fully evident and universal solution of the problem (A 716, B 744).

In order to avoid any misunderstanding of this approach Kant at once points out that construction is not identical here with geometrical drawing, and that algebraic procedure must also be regarded as construction, though of a symbolic kind. Operating with symbols in processes such as addition and subtraction corresponds to the drawing of straight lines in geometry (A 717, B 745). People have tried to interpret these remarks of Kant to mean that intuition is used in geometry as an additional help; but in reality being tied in this way to construction means a limitation. Of all possible proofs only those are to be admitted which can be supported by a construction.

That construction means a limitation, and not an aid, is made even clearer in the problem of definition. In the chapter on the distinction between phenomena and noumena Kant says, 'Real explanation would be that which made clear not only a concept but also its objective reality. Mathematical explanations, which exhibit the object in intuition in accordance with the concept, are of the latter kind' (A 242*n*.). Construction in intuition is therefore not an aid to definition, but limits the definitions which are possible in themselves to those whose concepts have objective reality. That is why the 'exactness' of the construction is quite unimportant. Even in the case of a circle it does not matter whether I draw an exact circle, and no mathematician would ever be able to or want to construct a chiliagon. But we have at our disposal the schema of the construction. In the polemic against Eberhard (VIII,192), Kant returns once again to this problem of construction and treats it at length. He explains there that mechanical construction is not intended; it does not matter at all whether I know how to construct a parabola mechanically. I only need to construct it schematically, and in some circumstances it can even be constructed analytically by means of an equation. It is quite clear to Kant that this kind of analytical construction by means of algebraic expressions has come to the fore in modern times; he is only pointing out that the synthetic method of the ancients by which they defined curves by means of concrete constructions of them should not be entirely forgotten.

The significance of what Kant says about construction only came to light through modern enquiries into the foundations of mathematics. It was Brouwer, the founder of the intuitionist school, who gave us genuine understanding of these problems. I will first give a short example. The Euclidean proof of the existence of an infinite number of prime numbers (*Elem.* ix.20) is a constructive proof. It not only proves the existence of a further prime number but also provides the means of actually calculating such a prime number or, in more general terms, of constructing it. The theorem that the continuum can be well-ordered is on the contrary of quite a different kind. It follows from the axiom of choice that any set is capable of being well-ordered.

This holds for example for the set of all real numbers. The proof however does not provide a means of making such an order. Here the schools of fundamental theory divide. The intuitionist school refuses to allow a proof such as that of the well-ordered continuum. This school only admits such concepts as can actually be constructed. In opposition to the intuitionists, other schools affirm the existence of mathematical objects if these objects are no more than free from contradiction. The well-ordered set of all real numbers is—as far as we know—free from contradiction, but cannot be constructed. Its existence is therefore denied by the intuitionistic school and affirmed by other schools. It is therefore plain that the principle of construction does not provide an additional aid for making proofs but selects some out of the number of logically possible concepts. Of all logically possible concepts, that is to say, concepts free from contradiction, only those are to be regarded as mathematically possible which can be constructed.

If we try to understand Kant's explanation of the constructive character of mathematics from this point of view, we are of course making use of things which Kant did not yet know in this precise way. It seems to me that we can use our present knowledge to explain Kant because the intuitionists themselves affirm their connection with Kant.[1] Thus the Kantian thesis that mathematics is intuitive in character means that mathematics is limited to objects which can be constructed.[2]

We are now in a position to clarify Kant's own attitude to Euclidean geometry. As we have already said, many Kantians have vigorously disputed the possibility of non-Euclidean geometry. This protest certainly has some ground in statements made by Kant, but the question is much more difficult than was at first assumed. It is made even more difficult by the fact that Kant—like Gauss later—avoided talking about non-Euclidean geometry, and in view of the conflicts that were kindled by the introduction of the non-Euclidean geometries, we have to say that Kant was right to be careful. But there can be no doubt that it was clear to Kant that in geometry the field of what is logically possible extends far beyond that of Euclidean geometry. There was, however, one thesis to which Kant held firm—presumably in error. What goes beyond Euclidean geometry may be logically possible, but it cannot be constructed, that is to say, cannot be constructed in intuition, and this in its turn means for Kant that it does not exist mathematically, that it is a mere figment of thought. Only Euclidean geometry exists in the mathematical sense and all non-Euclidean geometries are mere figments of thought. Kant gives a concrete example in his discussion of the Postulates.

> Thus there is no contradiction in the concept of a figure which is enclosed within two straight lines, since the concepts of the two straight lines and of their intersection contain no negation of a figure. The impossibility arises not from the concept in itself, but from the construction of the concept in space, that is, from the conditions and determination of space. But these have their own objective reality, that is, they apply to possible things, since they contain in themselves a priori the form of experience in general (A 220, B 268).

In terms of our present-day mathematical concepts this example is immediately intelligible to us. In Euclidean geometry two straight lines can never enclose a figure. But if the surface of a sphere is regarded as the model for a Riemannian geometry, then the great circles are Riemannian straight lines, and two straight lines can in fact make a figure (a lozenge). Yet until the question has been more closely investigated it must remain doubtful in what sense Kant envisaged a positive possibility of this kind. We know that Lambert saw the connection between non-Euclidean geometry and the surface of a sphere,[3] and it would not be unthinkable for Kant to have taken this train of thought a stage further. But it is also quite thinkable that Kant proceeded in a purely conceptual way and considered whether a contradiction is contained in the pure concept of a closed figure composed of two straight sides as such. At all events, Kant reached the correct view in one way or the other and saw that the concept is not a contradictory concept. By this approach Kant took a bold step beyond Leibniz, since for Leibniz the concept of such a figure is contradictory.

Kant thus denies the mathematical existence of a closed figure composed of two straight sides; it is indeed logically possible because it is free from contradiction, but it cannot be constructed and given in intuition. Kant's conception of non-Euclidean geometries in general has to be understood in the same way. Non-Euclidean geometries are logically possible but they cannot be constructed; hence they have no mathematical existence for Kant and are mere figments of thought.

This Kantian position could not be expressed in a precise form until the advent of our present-day knowledge—what Kant says of Leibniz is also true of himself, namely that later interpreters can understand a thinker better than he understood himself. But his attitude is perhaps not bold enough. It seems to be the case that there are ways in which non-Euclidean geometries can also be constructed, by purely analytical means or by constructing Euclidean models of non-Euclidean geometries, and certainly Kant's assumption that non-Euclidean geometries could never be used in physics has proved to be too narrow. On the other hand we should have to formulate the problem differently today and ask whether Euclidean geometry does not have a feature which distinguishes it from all other geometries, and this is a problem which we are as yet quite unable to solve. We know that there are Euclidean models of non-Euclidean geometries but we cannot yet survey the connections between all possible geometries. Is one geometry perhaps, and presumably then Euclidean geometry, distinguished by its logical form, or perhaps distinguished by being the foundation of the other geometries, or by being the only way through which we can reach the other geometries? Or is no particular geometry distinguished in this way? We are only mentioning these questions to bring out the significance of Kant's thesis. When Kant emphasises the importance of intuition for geometry he means to say, if we are right, that many geometries are thinkable without contradiction but that from the extensive region of what is thinkable without contradiction Euclidean geometry is singled out and distinguished by the fact that it can be constructed. It is then merely a question of terminology whether something that goes beyond Euclidean geometry, but only as thinkable without contradiction, should still be called a geometry. Kant would probably have hesitated to give the name of a geometry to what he calls pure figments of thought. But as the usage of the term has developed in the way it has done I can see no objection in principle to using it of Kant. The Kantian example of a two-sided figure that we have been discussing seems to be in favour of this usage, and every interpretation has to work with concepts that have undergone further development if it is not to be mere repetition.

Intuition, then, is not an additional source of knowledge for mathematics—this thesis has been rightly rejected by mathematicians—but is the factor which limits the broader region of logical existence, namely what is thinkable without contradiction, to the narrower region of mathematical existence, namely what can be constructed. This is the second respect in which geometrical judgments are synthetic judgments a priori, namely because they can be constructed in pure intuition.

In thus exhibiting the constructive character of geometry a further element of spontaneity is brought to light.[4] Mathematical thinking is not merely thinking with pure concepts, it is also constructing; nowadays we should perhaps say in rather more general terms, it is operating. In the passage quoted above Kant explains why he regards geometrical proofs as constructive proofs. When the geometer wants to prove the sum of the angles of a triangle, he begins, no doubt, by drawing certain lines; that is simply a description of the facts. But the same also holds in arithmetic. The example given in the Introduction to the second edition of the *Critique of Pure Reason* is perhaps not altogether happy. Kant says:

> The concept of 12 is by no means already thought in merely thinking this union of 7 and 5; and I may analyse my concept of such a possible sum as long as I please, still I shall never find the 12 in it. We have to go outside these concepts and call in the aid of the intuition which corresponds to one of them, our five fingers for instance, or, as Segner does in his Arithmetic, five points, adding to the concept of seven, unit by unit, the five given in intuition. For starting with the number 7, and for the concept of 5 calling in the aid of the fingers of my hand as intuition, I now add one by one to the number 7 the units which I previously took together to form the number 5 and with the aid of that figure see the number 12 come into being (B 15).

It is understandable that mathematicians have not been very pleased about this reduction of arithmetical proofs to the five fingers, and the reference to Segner's textbook has not helped to reconcile them to it. But it can be shown that there is an exact mathematical basis for this example. Kant conceived the proof in pure mathematics in the following way:

$$7+5=7+(4+1)=7+(1+4)=(7+1)+4=8+4=\text{etc.}$$

The proof proceeds by taking the units one by one from the 5 over to the 7; and this is how Kant actually describes the procedure in his vivid account of it.[5] Leibniz had already seen that this was a fundamental proof and had given an account of it in the *New Essays* (IV.7.10). The constructive character of this proof is grounded in the fact that repeated transformations have to be made. The 5 becomes $4 + 1$ by definition, but then the commutative law has to be applied to transform this expression into $1 + 4$, and then again the associative law has to be applied to transform the expression $7 + (1 + 4)$ into $(7 + 1) + 4$, and only then can one go by definition from $7 + 1$ to 8. Arithmetical proofs rest on such transformations and are therefore constructive.[6] To build up an arithmetical proof means actually to set up these transformations and to carry them out; thus not only every geometrical proof but also every arithmetical proof is an *operari*. Kant believed that it was better to develop this in pure intuition by means of the five fingers or five points; nowadays the purely arithmetical presentation of the problem is perhaps more convincing.

To show successfully the constructive character of geometry, indeed of mathematics as a whole, is thus to bring out that mathematical existence implies constructability and that mathematics is operative in character. At the same time this provides a new factor in support of the synthetic character of the science of mathematics.[7]

§ 4

The *a priori* Character of Geometry

Our results so far show that space is necessary; it is not necessary in thought but only has the peculiar necessity of a mathematical existence grounded in construction in intuition. To be exact, this necessity holds of Euclidean space. This necessity, which issues for Kant in the problem of the a priori character of space, is the foundation for his further thesis that space is an appearance. This thesis is very closely connected with the definition of space that Leibniz had given. We saw that Leibniz brings together a thesis in categorial analysis and one in ontology; from the point of view of categorial analysis space is seen to be a mere complex of relations, and being a relation, space must be phenomenal, since relations have no existence of their own but have their origin in the understanding (G.II,517).

As regards the first aspect of space, that of categorial analysis, Kant retains Leibniz's results; for Kant also, space is merely a complex of relations. In the Note to the Amphiboly of the Concepts of Reflection (A 284, B 340) Kant speaks of 'a space which with all that it contains consists solely of relations, formal or real'. Space consists only of relations and that is why, in the survey of existence as a whole in the Transcendental Aesthetic from which we started, space and time are regarded from the first as relations. Kant mentions but does not seriously discuss the possibility that they might be substances or accidents, and the question at issue is solely whether they represent relations of things in themselves or whether they are relations which have their origin solely in the knowing and intuiting mind.

In the additions to the Transcendental Aesthetic in the second edition, Kant alludes explicitly to the relational character of space and time.

In confirmation of this theory of the ideality of both outer and inner sense, and therefore of all objects of the senses, as mere appearances, it is especially relevant to observe that everything in our knowledge which belongs to intuition contains nothing but mere relations (B 66).

As regards the ontological status of space, Kant again accepts Leibniz's definition in his early writings. Space is phenomenal; this ontological thesis is taken over by Kant, just as is the thesis in categorial analysis that space is relational in character. The truth of this statement is not affected by Kant's transitory inclination towards Newton in the short essay written in 1768 with the title *Of the First Ground of the Distinction of Regions in Space*. But although Kant kept to the principle of defining space and time as phenomenal, it is not easy to say whether he followed out this approach to its final consequences and whether he defined space and time in the final formulation of his theory, in the same way as Leibniz, as *phaenomena Dei*. Kant may well have found difficulties in the transcendental theology of the proposition that we see all things in God, which is linked with this definition, and which he mentions at the end of Section IV of the *Dissertation* of 1770 (II,410). He may also have met quite early the difficulties that this produces in theory of science. The definition of space and time as *phaenomena Dei* rests on the uniqueness of space and geometry; three-dimensional Euclidean geometry has to be the only geometry that is possible without contradiction, and Kant had doubts early in his life about this assumption. But in whatever way Kant's thought may in fact have developed, this is the point at which he diverges from Leibniz. Kant keeps the denial of the absolute reality of space as substance or as accident, he keeps the relational character of space, he keeps the merely phenomenal existence of space. The point at which the divergence begins is when Leibniz tries to refer the phenomenal character of space to God's reason. This leads to the decisive change-over; the thought for which space is an appearance is now human thought and Kant thus arrives at the celebrated thesis, 'We can therefore only speak of space, of extended beings, etc. from the standpoint of a man' (A 26, B 42).

The first two arguments about space are especially important for supporting this thesis.

In the first argument about space, what must first be noticed is that it is taken over almost word for word from the *Dissertation*. The first argument aims to prove the a priori character of the representation of space. It consists of three sentences which make up the three parts of the argument. In the first sentence, 'Space is not an empirical concept which has been drawn from outer experiences', the aim of the argument is indicated. The second sentence gives the actual proof. 'For in order that certain sensations be referred to something outside me (that is, to something in another place in space from that in which I find myself), and similarly in order that I may be able to represent them as outside and alongside one another, and accordingly not only as different but as in different places, there must already be an underlying representation of space.' The third sentence draws the conclusion. 'The representation of space cannot, therefore, be empirically obtained from the relations of outer appearance; on the contrary this outer experience is itself possible at all only through that representation' (A 23, B 38).

The kernel of the argument lies in the proof given in the middle sentence. Kant speaks here in terms of sensations and the question is whether the concept of sensations as such is essential. That this is not the case can be shown from the *Dissertation* (§ 15A), where Kant only speaks of *concipere*.

For I cannot conceive (*concipere*) anything as placed outside me unless by representing it as in a place different from that in which I myself am; nor can I conceive things as outside one another unless by locating them in different places in space (II,402).

If we substitute for 'sensation' the general term 'something', we can then extract the pure kernel of the argument in the following way: in order for me to be able to represent something as alongside something else, there must already be an underlying representation of alongside.

There is the further question as to whether this argument is an argument specifically about space and whether Kant is here making use of a specific property of space. Is the argument limited to the representation of space or could it perhaps be extended to include other representations? The argument aims at establishing the a priori character of space, and we must therefore ask whether the argument establishes an a priori character solely for space or whether the a priori character of other representations can also be established by it. Is it perhaps even an argument by means of which the a priori character of every a priori concept can be established universally? This is in fact the case. One only has to look at the treatment of time in the Transcendental Aesthetic to see that the argument is valid at least for the representation of time, since the first argument about time is almost word for word identical with the first argument about space. If one pursues the matter one soon sees that the argument goes very much farther, that it proves the a priori character not merely of pure intuitions but of all a priori representations, in so far as it is possible to prove their a priori character at all. The argument is in fact Plato's fundamental argument, and for this reason I shall call the first argument the Platonic argument.

Plato uses this argument continually to provide a foundation for his doctrine of ideas. It is expounded in a particularly clear and lucid way in the *Phaedo* (74e). Plato explains his view here by means of the problem of equality. In the first instance we are given equal pieces of wood and equal stones, but the equal pieces of wood and equal stones do not exhaust the problem of equality. Beyond them there is equality as such, equality in itself. But where do we get our knowledge of equality as such? In order to be able to recognise two pieces of wood as equal at all, we have to know in advance what equality is. We certainly cannot have taken the concept of equality as such from the equal pieces of wood and the equal stones, because these are never really equal but only approximately so. In order to recognise two pieces of wood as equal at all we have to know in advance what equality is.

In order to be able to recognise two things as equal I have to know in advance what equal means—I must have the concept of equality at my disposal. In the same way Kant says that I can only recognise two things as alongside each other if I already have the concept of alongside, therefore there must already be an underlying concept of alongside.

The connection between the first argument and Plato's fundamental approach was noticed early. Natorp re-discovered the connection [1] and Hartmann later emphasised the roots of the Kantian a priori in the Platonic *προειδέναι*.[2] The historical importance of the argument is in fact as great as its systematic importance. Plato wants to prove the a priori character of ideas. He first succeeds in expounding the theory systematically for a mathematical subject, in the *Meno*. A slave is called. Through Socrates's questions alone the slave discovers that the square on the diagonal of a square is double the size of the square itself, a proposition that played an important part in Greek mathematics in Plato's time because of its connection with Pythagoras's theorem and the problem of irrationals. The slave, as his master is able to assure us, had never studied mathematics before and Socrates did not tell him anything but only asked him questions. From this Plato infers that the slave drew this geometrical discovery from out of himself, and in general that we always draw mathematics from out of ourselves. Plato regards all knowledge of ideas, in this case knowledge of mathematics, as recognition and as reminiscence. It is not perhaps out of the question that reminiscence is a myth for Plato and that he is trying to present considerably more abstract matters in the form of this myth. There can be no doubt that, for all that, Plato grasped something fundamental and irrefutable. The mathematician's knowledge is gained in a different way from what is known in everyday life or in the experimental sciences. This is something irrefutable, although it can perhaps be interpreted in different ways. It was the starting-point for Plato, Aristotle, Thomas Aquinas, Leibniz, and Kant, even though they interpreted it differently. Common to all these thinkers is the conviction of the a priori character of mathematics, and their differences lie only in their interpretation of the a priori.

It would be attractive to develop the underlying problem in Plato's argument historically. It would come out more and more clearly that the argument extends far beyond the region of mathematics and that in Plato, as later, it is used as a foundation for the a priori character of pure concepts in general. Even for Plato the argument holds not only for the problem of doubling the square and the problem of equality; it also holds in the same way, for example, for the concept of water. In order to recognise something as water and in order merely to be able to call something water, I have to know in advance what water is. In order to subsume something under the concept of water, I must have in advance the concept of water, in Plato's language the idea of water, and the argument is used in this sense in the *Timaeus*. We are at present only interested in its significance for mathematics, and the argument proves for Plato as for Kant the a priori character of all mathematical concepts; in order to recognise something as alongside and in order to be able merely to represent anything as alongside at all, I have to have the representation of alongside in advance.

If the first argument is based on the Platonic doctrine of ideas, the second argument is based on the Aristotelian ontology. I shall therefore call this the Aristotelian argument. A history of the problem that is involved here could also be very attractive, but we shall have to be satisfied with showing its connection with the form the argument took for Kant. It again consists of three parts each expressed in a sentence. The first sentence gives the aim of the argument. 'Space is a necessary representation a priori which underlies all outer intuitions.' The second sentence contains the argument proper. 'We can never represent to ourselves the absence of space though we can quite well think it as empty of objects.' The third sentence again sums up the result. 'It must therefore be regarded as the condition of the possibility of appearances, not as a determination dependent on them, and is an a priori representation which necessarily underlies outer appearances' (A 24, B 38).

The kernel of the argument is again in the middle sentence. One can think space without objects but one can never think objects without space, and it follows from this that space has precedence over spatial objects. This is the fundamental inference, and it takes up both a specifically spatial and a general problem of Aristotelian ontology.

Aristotle treats the problem in the fourth book of the *Physics*, in the discussion of space and time. The first of the five chapters on space is devoted to the history and the aporetic of the problem of space. Space had been called a reality by the very early thinkers; Aristotle mentions Hesiod. The problem of space then became especially important for the atomists. Anyone who assumes the possibility of the void, as do those who hold atomistic theories, has to attribute independent existence to space, because an empty space is after all a space out of which the objects have been removed, which has been robbed of its objects. The conception of space as something existing that precedes everything else that exists is in Aristotle's belief much older than this. Aristotle identifies Hesiod's chaos with space. When Hesiod says that chaos was of all things the first to come into being, this means that space is something that existed before all other things. If this were the case, as Aristotle comments at once, the nature of space would be indeed wonderful and space would have precedence over all other things. Aristotle then proceeds with his ontological argument in the following way (*Physics* IV.1.208a27): if there is something without which all other things cannot be and which itself can be without other things, this is necessarily the first and original thing. This ontological relation does in fact hold of space. Things in space come into being and pass away, but space itself does not participate in this coming to be and passing away, but subsists continuously and immutably. This argument springs from the fundamentally competitive attitude of the Greeks. Two Greeks cannot meet without comparing themselves with each other to find out who is the first and two Greek towns cannot help finding out at once which town enjoys precedence. Things are regarded from the first with a view to their possible order of precedence. Space precedes spatial things. This relation of precedence then takes on a special form in the relation of substance and accident.

The same relation as holds between space and spatial things also holds substance and accidents. There can never be the wrath of Achilles without Achilles, but there can be Achilles without the wrath. In general terms substance is that which always exists with or without particular accidents while the accidents cannot exist without the substance. The atomistic doctrine leads, as we have already seen, to space being described as a substance, and Aristotle emphasises sharply the definition of space that leads to this conclusion, namely that of being something other than bodies.

This mutual relation between space and bodies need not necessarily lead to the substantification of space, but it always tends to give a precedence to space. This is the case even when the argument appears in a different form, as with Kant, although of course the change had taken place very much earlier. In his discussion of space and bodies, Kant only speaks of their being represented—I can represent to myself space without bodies but not bodies without space. But this alteration to the argument does not affect the outcome. It simply transforms an actual experiment, the removing of bodies from space, into an experiment in thought. It may not be possible actually to remove bodies out of space but I can still carry out this experiment in thought, and the experiment in thought in fact produces the same result. In scholasticism, where the alteration to the argument first appears, a middle solution is at first found. It may be the case that bodies cannot really be taken out of space, but we can still ask whether God's omnipotence could not remove them and whether perhaps, at least through divine intervention, space might exist without bodies. If space existed before the creation of the world then there has actually been space without bodies. The Aristotelian argument is then very important for Leibniz. For Leibniz monads can exist without space but space cannot exist without monads. If monads were destroyed, for instance by divine intervention, space would also be destroyed (G.II,510, to Des Bosses). If there were no created things there would be neither time nor place nor any actual space (5th Letter to Clarke, § 106). This argument about the relation of space to spatial things runs like a crimson thread through the whole discussion of space, so that it is not surprising that Kant also uses it. The argument proves conclusively for Kant that space precedes spatial things because, at least in the experiment in thought, space can exist without things, but not things without space.

Thus the two arguments, the Platonic argument and Aristotelian argument, come together for Kant and prove definitively the precedence of pure space and pure time. He expresses this precedence in the form of the thesis that space and time are a priori in character.

§ 5

Geometry and Natural Science

Up to now we have regarded space as a purely mathematical phenomenon. We were able to consider the axiomatic, the constructive, and the a priori characters of space without regarding space and geometry as anything other than a purely mathematical phenomenon. But mathematics, besides existing as pure mathematics, is also applied to nature, and even if the pure mathematician considers this to be nothing more than a mere application, for the natural sciences the relation is fundamental. The question now is, what is the significance for natural science on the one hand, and for mathematics on the other, of the application of mathematics to nature.

There can be no significant natural science without mathematics, but can there be significant mathematics without its application in natural science? Kant's answer is no. Mathematics without a possible application is meaningless. One is inclined to reject this thesis out of hand, but yet Kant could claim that nearly all those who have reflected on the meaning of mathematics have agreed with him.

One might at first expect that Plato would have rejected such a thesis completely, and there is a form of the doctrine of ideas in which it is quite indifferent to the ideas whether they have any connection with the world. But if we may follow Paul Natorp's interpretation, Plato recognised more and more deeply that ideas do not exist by themselves but are connected with each other and with the world in the closest possible way. This thesis is expounded in the late form (the so-called dialectical form) of the doctrine of ideas in the *Parmenides* and the *Sophist*. Natorp seems to me to have shown that the ideas determine the existence of the world.[1] The thesis is developed systematically by Aristotle. Mathematical laws are for him the laws of existence and thus the laws of nature, and in a certain sense Aristotle can be said to have regarded mathematics as a natural science. But mathematics treats the plainest and simplest matters, so that here again it occupies a very special position (*Met.* XIII. 2.1077b12). In Leibniz we find, as so often, a surprising combination of the Platonic and Aristotelian approaches. Geometry exists on the one hand as an ideal structure of what is possible without contradiction and on the other hand as the structure of the actual world and of every possible world. This line of thought finds its highest form of expression in the assumption which we have already considered, namely that mathematics is to be referred to God's reason, and thus that mathematics consists primarily in God's seeing the world mathematically.

It is interesting to note in this connection that Bertrand Russell, in the first edition (1919) only of his *Introduction to Mathematical Philosophy*, went so far to call mathematics a natural science.[2] From an ontological approach Nicolai Hartmann has also emphasised that ideal being must not be regarded solely in itself, and that ideal being as such has a connection with real being. He maintains that this is not inconsistent with the transcending of real being by ideal being. Hence mathematical existence also transcends the possibilities of real existence but in spite of this retains its original reference to the real world.[3]

This reference of mathematics to the real world is specially emphasised by Kant.

Therefore all concepts, and with them all principles, even such as are possible a priori, relate to empirical intuitions, that is, to the data for possible experience. Apart from this relation they have no objective validity and are a mere play of the imagination or of the understanding each with their representations. Take, for instance, the concepts of mathematics, considering them first in their pure intuitions. Space has three dimensions, between two points there can be only one straight line, etc. Although all these principles and the representation of the object with which this science occupies itself are generated in the mind completely a priori, they would mean nothing at all if it were not that we are always able to use appearances (empirical objects) to exhibit their meaning (A 239, B 298).

Mathematical concepts that cannot be referred to empirical intuitions are thus empty, they are mere play of the imagination. This thesis means in concrete terms that non-Euclidean geometries are without meaning, notwithstanding their being possible without contradiction, and that they have no existence either as applied mathematics or even as pure mathematics. For Kant the ideal possibilities of mathematics do not in principle go beyond the possibilities of real existence. This emerges particularly clearly if Kant's doctrine is compared with that of N. Hartmann.[4]

Kant is nevertheless able, through placing this restriction upon mathematics, to clarify the problem that is involved. The question is, why Euclidean geometry, in spite of being a pure construction in thought, is valid for Newtonian physics. Through the restriction of mathematics to what can be constructed, the solution to the problem takes the form of showing that the physicist makes constructions in fundamentally the same way as the mathematician.

The two previous principles, which, in view of their justifying the application of mathematics to appearances, I entitled the mathematical, referred to the mere possibility of appearances, and taught how, alike as regards their intuition and the real in their perception, they can be generated according to rules of a mathematical synthesis. Hence both principles justify us in employing numerical magnitudes and so enable us to determine appearance as magnitude. For instance, by combining some 200,000 illuminations by the moon, I can give the degree of sensations of sunlight as determined a priori, that is, I can construct it (A 178, B 221).

The same fundamental act, that of construction, which generates the objects of mathematics also generates the objects of natural science, so that the validity of geometry in physics is thereby directly given.

By discovering the connection between mathematics and natural science Kant fundamentally went far beyond the concept of applied mathematics. What we have is not a ready-made mathematics which is applied in physics, but one fundamental human faculty which is active in mathematics and in physics; not an application of a ready-made mathematics, but an original spontaneity reaching out into further regions. The concept of spontaneity and thus the concept of synthesis now receive a new and extended significance. The spontaneity of pure knowledge generates not merely the objects of mathematics but goes further and generates the objects of natural science.

§ 6

THE TRANSCENDENTAL IDEALITY OF SPACE AND TIME

Kant epitomises his views on the ontological status of space and time in the proposition that they are transcendentally ideal and empirically real. He says of space, 'We assert therefore the empirical reality of space . . . yet at the same time its transcendental ideality' (A 28, B 44), and again of time, 'What we assert is therefore the empirical reality of time . . . but on the contrary we deny all claim to the absolute reality of time . . . and that is what transcendental ideality consists of' (A 35, B 52).

Kant defines the ideality of space and time as a transcendental ideality. Here there is a difficulty, because in the term 'transcendental' a meaning which comes from the ontological tradition is combined with a new and specifically Kantian meaning. It can be shown that the meaning which comes from the ontological tradition is still alive for Kant. He speaks in this sense of a transcendental philosophy of the ancients (B 113). The term 'transcendental' is found in the pre-critical writings, even in the very early ones.[1]

The connection is particularly clear when Kant is discussing absolute and transcendental reality (A 36, B 53). In this passage the term certainly cannot have its specifically Kantian meaning, for he says explicitly that the theory developed in the *Critique of Pure Reason* denies that time has absolute and transcendental reality; it denies that space and time are things real in themselves. Here, therefore, the term must have its traditional meaning. 'Transcendental' denotes here the most general determinations of being, namely *ens*, *unum*, *bonum*, *verum*, both as regards the object and as regards method. A transcendental enquiry is one that only considers its objects in respect of these universal determinations, and a transcendental theology, for instance, is a theology that is only concerned with the most universal determinations of God and not, for example, with the problems of the trinity or of christology. Kant himself uses the term 'transcendental theology' in this sense (A 580, B 608).[2] When Kant speaks of transcendental ideality or reality he means therefore that the existent in question is ideal or real considered in respect of its universal ontological character. If we try to express Leibniz's fundamental ontology in this Kantian language, transcendental reality would have to be attributed to monads and transcendental ideality to space and time. The proposition that space and time are transcendentally ideal means therefore that space and time in respect of their being as such are not real but ideal.

The specifically Kantian meaning of the term 'transcendental' is only possible on the basis of the traditional ontological meaning. In the Introduction to the first edition Kant gives the following definition, 'I entitle transcendental all knowledge which is occupied not with objects but with our concepts a priori of objects in general' (A 11). In the second edition he gives a more precise definition: 'I entitle transcendental all knowledge which is occupied not with objects but with the mode of our knowledge of objects in general in so far as this mode of knowledge is to be possible a priori' (B 25).

Kant therefore asserts first the transcendental ideality of space and time, and this thesis is one that he shares with Leibniz.

He then asserts further that the transcendental ideality of space and time has its origin in the nature of the human mode of knowledge, and with this assertion he parts company with Leibniz. From the Kantian point of view a transcendental enquiry into the transcendental ideality of space and time must be directed primarily to the origin of space and time; that is to say, from the Kantian point of view a transcendental enquiry must be an enquiry into the human mode of knowledge (B 25).

The transcendental ideality of space and time therefore implies not only the denial of their absolute reality but also the further thesis that the origin of space and time as ideal beings lies in the human mode of knowledge. Kant summarises his thesis accordingly.

> What we have meant to say is that all our intuition is nothing but the representation of appearance; that the things which we intuit are not in themselves what we intuit them as being, nor are their relations in themselves as they appear to us, and that if we as subject, or even only the subjective constitution of the senses in general, be removed . . . all the relations of objects in space and time, nay space and time themselves, would vanish; as appearances, they cannot exist in themselves, but only in us (A 42, B 59).

Kant had already reached this point of view in the *Inaugural Dissertation*. There he calls time 'a subjective condition necessarily dependent on the nature of the human mind (§ 14,5; II,400), and in the same way space is also called 'a subjective condition'. Space and time are thus now regarded as human conditions. Kant expressed this result in the impressive thesis that 'we can therefore only speak of space, of extended beings, etc., from the standpoint of a man' (A 26, B 24). To summarise again what we have said in the foregoing paragraphs, for Kant space means Euclidean space, and in particular Euclidean geometry. Both Kant and Leibniz assert the objective validity of Euclidean geometry. Leibniz grounds the objective validity of Euclidean geometry on God's thinking, Kant grounds it on human thinking. The transcendental ideality of Euclidean space then means that the being of Euclidean space consists in its being thought, for Leibniz in its being thought by God, for Kant for its being thought by man. This, that the being of Euclidean space consists of its being thought by man, is what Kant means by the transcendental ideality of space.

In addition to the term 'transcendental ideality' Kant also uses the term 'appearance' to characterise space and time ontologically. As the *Critique of Pure Reason* proceeds the term 'appearance' is in fact used predominantly.

The thesis that space has the character of appearance had also been taken over by Kant from Leibniz; Leibniz had described space and time as phenomena (G.II,435). Wholly in Leibniz's sense Kant says in the *Monadologia physica* of 1756, 'Space is not a substance, but is a certain phenomenon of the external relations of substances' (I,480). In the *Dissertation* again the phenomenal character of space is emphasised. Kant says of space 'this concept contains the universal form of phenomena' (II,401) and the same of time, 'a certain subjective condition of all phenomena' (II,404).

The definition given in the Transcendental Aesthetic is, 'Space is nothing other than the form of all appearances of the outer senses' (A 26, B 42), and similarly of time, 'Time is the formal condition a priori of all appearances in general' (A 34, B 50).

This thesis is of course a double thesis, since space and time are immediately further defined as forms of our intuition. Space and time 'taken together, are pure forms of all sensible intuition' (A 39, B 56). This is where the so-called identity-theory, to which Nicolai Hartmann has so insistently drawn attention,[3] first becomes effective. Space and time are at the same time forms of appearances and forms of sensible intuition. There has been much controversy (fully reported in Vaihinger's commentary [4]) as to whether Kant's assertions refer to pure or to applied mathematics. It seems to me that this dispute is fundamentally mistaken and that Kant is concerned both with pure and with applied mathematics.

To go back to what we have already said, we have shown that for Kant other spaces are entirely thinkable but that, in the full sense of mathematical existence which includes for Kant constructability, only Euclidean space exists. Only Euclidean space can be constructed. It follows from this for pure mathematics that every geometry is Euclidean. We have already considered in what sense the development of mathematics has gone beyond Kant and in what sense the question is none the less still completely open.

From this approach the Euclidean character of physical space also follows. We assert once more that Kant maintains both the Euclidean character of mathematical space and the Euclidean character of physical space. From this in turn two things follow: the proposition that Euclidean geometry has objective validity, and the proposition that all spatial determinations are limited to appearances.

The proposition that Euclidean geometry has objective validity is enunciated by Kant again and again. For Leibniz mathematical existence means freedom from contradiction and Leibniz therefore has to prove that the axioms of Euclidean geometry are free from contradiction. In the sense of mathematical existence re-defined by Kant as constructability, there only is three dimensional Euclidean space. Leibniz and Kant go different ways but the result is the same: only three dimensional Euclidean space is possible.

But the objective validity of geometry is deduced by Kant from the identity-theory (that space and time are forms both of appearances and of sensible intuition) and it can therefore only hold within this theory. The objective validity of geometry can be proved by Kantian arguments only for appearances and not for things in themselves.

Leibniz could avoid this consequence in one sense, because for him the thinking that carries space and time is God's thinking. Kant removes this foundation by referring space and time only to human thinking and therefore has to remove finally all spatial properties from existence in itself.

It has often been noticed that there is a gap here in the Kantian proof.[5] But the gap is not in the proof but in the formulation. Kant should not really have said that spatial properties are only attributable to appearances and not to things in themselves. He should only have said, it does not necessarily follow from the fact that appearances necessarily have Euclidean character that things in themselves also have Euclidean character. He thus really ought only to have said, it need not follow from the fact that appearances are in space and time that things in themselves are also in space and time. We shall see that in the Transcendental Analytic Kant chooses the more cautious and precise formulation, but the problem of space and time is loaded with so much feeling that the emotional attitude carries the day over the cooler ontological formulation.

If we try to keep within the framework of what can be proved by the Kantian argument, we can say that it is possible to demonstrate the empirical reality of space and time, that is to say, the objective validity of all spatial and temporal properties in mathematics and physics. But this empirical reality involves transcendental ideality; space and time are forms of human intuition, and they can only be proved valid for things as they appear to us and not for things as they are in themselves.[6]

CHAPTER II

THE WORLD

§ 7

The Problem of Antinomies and its Historical Antecedents

DEPARTING from the order of the *Critique of Pure Reason* we now proceed from the Transcendental Aesthetic to the Transcendental Dialectic, leaving the Transcendental Analytic till third. The problems may be presumed to have developed in this order for Kant himself. The *Dissertation* shows clearly that by 1770 the standpoint of the Transcendental Aesthetic had been reached to the extent that important passages from the *Dissertation* could be taken over into the *Critique of Pure Reason* almost word for word. The standpoint of the Transcendental Dialectic had also been to a large extent reached. But as regards the problems of the Transcendental Analytic Kant takes up in the *Dissertation* the position which he later called dogmatism, and the standpoint of the Analytic must therefore have been reached in the years between 1770 and 1780. This inference from a comparison of the *Dissertation* with the *Critique* is confirmed by direct evidence contained in the long letter of 21 February 1772 to Marcus Herz (X.123). The problems of the Transcendental Dialectic therefore represent an important stage in the development of critical thought. Benno Erdmann drew special attention to these problems, and in 1938 de Vleeschauwer made them the subject of an enquiry, in an article on the Antinomies in *Mind* (N.S. 47, p. 303*ff.*).[1] We confine our running commentary to the doctrine of the Antinomies, where fundamental questions of reason are raised:

> Whether the world has a beginning in time and any limit to its extension in space; whether there is anywhere, and perhaps in my thinking self, an indivisible and indestructible unity, or nothing but what is divisible and transitory; whether I am free in my actions or, like other beings, am led by the thread of nature and of fate; whether finally there is a supreme cause of the world, or whether the things of nature and their order are the ultimate object, at which all our contemplation must come to a halt: these are questions

for the solution of which the mathematician would gladly exchange the whole of his science (A 463, B 491).

We shall confine ourselves further to discussing the first two Antinomies, since they enable us to show sufficiently clearly the peculiar way in which historical antecedents and systematic foundations are connected, and how this has important consequences in transcendental philosophy.

The questions that are dealt with in the first two Antinomies have been matters of philosophical concern from the earliest times. In the Transcendental Dialectic the problems are reduced to their pure systematic form, but Kant is well aware of the historical antecedents and repeatedly refers to them. To eighteenth-century readers they were familiar but at the present time a review of them seems to be called for.

The first Antinomy divides into the two questions as to whether the world is finite or infinite in space and finite or infinite in time. As regards the first question, whether the world is finite or infinite in space, in mythological thinking the world has always been conceived to be finite. The sphere of the heavens encloses the world and makes it in the proper sense finite. The idea of a world stretching to infinity, or as would be said mythologically, losing itself in infinity, is never seriously considered. This conception persists far into educated thinking. But since there is a final limit, existence is 'completed in all directions, like the mass of a well-rounded sphere', as Parmenides says (fr. 8). He can hardly be thinking here of an exclusively abstract concept of existence and must be thinking of existence as the cosmos itself. This combination of a simple picture and abstract thought, which is so difficult for us to imitate, is fully developed by Aristotle. In the Aristotelian cosmology the earth is a ball resting at the centre of the universe, surrounded by a series of spherical shells, first of water, then of air, and then of fire. The ether as the fifth element and the heaven of the fixed stars complete the universe in the form of a perfect sphere, and the possibilities of existence so far as it is capable of being moved unfold within this sphere and are confined to it.

The idea of a finite world is always predominant in Greek thought, but through the untiring versatility of Greek thinkers the other possibility, the idea of an infinite world, is not overlooked. The space of the atomists has to be thought as unlimited and as infinite. The space also that Plato describes in the *Timaeus* as the third principle no longer has the limited form of the sphere. The notion of an infinite universe reaches its purest form with the Pythagoreans. They make *peras*, the limited, and *apeiron*, the limitless, into principles (Aristotle, *Met.* I.5.986a17), and there can be no doubt that space belongs to the side of *apeiron*. This is most clearly expressed in the fragment of Archytas (A 24) quoted by Eudemos. If I stand at the outermost edge of heaven, at the place where all change is left behind, can I stretch out my hand or my staff into that which lies outside it, or can I not?

Here the aporia in the idea of a limit to the world is brought out with a sure touch. If there were a limit to the world, I could walk up to this limit and still stretch out my hand, but then I should be going beyond the limit into that which must lie beyond it. I must therefore be able to step beyond any limit to the world, wherever it may lie, and this is the unresolvable antinomy of a finite limit to the world.

In the early stages of modern natural science, the idea of an infinite world comes so far to the fore that a finite world is no longer considered, even as a possibility. For both the theories which then develop, space as such is infinite. Newton assumes a finite world situated in a space which is infinite in itself and otherwise empty. Leibniz on the other hand holds this kind of a world, a finite island in infinite space, to be impossible, and for him space is continuously filled with matter to infinity (4th Letter to Clarke § 21, and Clarke's 3rd Reply, § 9). This is the form in which the conflict appears in the correspondence between Leibniz and Clarke, and in which it must have exerted its influence on Kant. Kant thus takes over in this restricted form the problem of a finite world in infinite space or an infinite extension of matter into infinite space.

The question of a beginning of the world in time played perhaps an even more important part. In this problem as in that of space, thinking starts from a finite concept, the beginning of the world at a finite time. Practically all mythologies begin with the story of a creation. The world is created by God, or by Gods, or by beings which were created for this purpose by the Gods. This kind of a myth of creation is given by Plato in the *Timaeus* (27c*ff.*). The world is said to be created after the pattern of the ideas, not by God himself, but by a world-builder, the Demiurge, created by God. There was controversy among Plato's pupils as to whether the account of the creation in the *Timaeus* was an account of a real event in time and was thus intended to represent a real temporal beginning of the world, or whether it was only a myth which was not intended to be believed literally. According to the account of the creation in *Genesis* God created heaven and earth in the beginning, but not all interpreters are of the opinion that this 'in the beginning' is to be taken directly in a temporal and finite sense. But in general the account in the *Timaeus* and the story of the creation in *Genesis* have been taken as descriptions of an event that happened at a finite past time, and the agreement between the Platonic account and the story in *Genesis* has very important consequences.

In opposition to the thesis that the world had a beginning at a finite time, Aristotle upholds the doctrine of a world that is infinite in duration. He relies primarily on the motion of the stars. From the geocentric point of view the fundamental motion appears to be the rotation of the heaven of the fixed stars, which revolves once in twenty-four hours. The motions of the sun, the moon, and the planets are then derived from this fundamental motion by being regarded as composed of several circular motions. The whole system moves changelessly, and it is the rotation of the fixed stars especially which goes on unchanged from all eternity to all eternity. It is thus the constant motions of the celestial bodies which constitute for Aristotle infinity in time.

This disagreement between Platonic and Aristotelian doctrine became a problem when Aquinas made the momentous transition from Plato to Aristotle. The Platonic cosmology is in agreement with the story of the creation, whereas the Aristotelian cosmology contradicts at least the wording of *Genesis*. The question as to what Aquinas's attitude to this problem actually was is not without its difficulties and has led to much controversy between his interpreters. We may presume that it cannot be definitively clarified. In Aquinas's view, as I read him,[2] Plato and Aristotle are not discussing the real duration of the world but only pure possibilities. The problem can only be formulated from the standpoint of pure possibility, since we have no experience of the beginning of the world and can have none. From the standpoint of pure possibility both theses are equally valid in logic, or in concrete terms God could have created the world either at a finite past time or at an infinite distance in time. The existence of the world from an infinite time is not in contradiction with its character as created existent, as *creatura*, and it is in this character alone, namely as a created being, that the world can be grasped in pure knowledge. But we know from the Bible that the world was in fact created at a finite time; this fact is a revealed truth. The question cannot therefore be decided by means of natural reason and in particular cannot be decided by philosophers. From this point of view the conflict between Plato and Aristotle is thoroughly understandable.

The problem of the beginning of the world in time also plays a large part in the correspondence between Leibniz and Clarke, and is thus in dispute between Leibniz and Newton. Newton stands for the creation of the world at a finite time, while conceiving time itself to be infinite. Leibniz on the contrary stands for an infinite duration of the world, but he also considers the possibility of a finite duration of the world and can therefore be said in a sense to have reckoned with the antinomic character of the problem (5th Letter, § 56). We found that Aquinas had already understood the conflict between Plato and Aristotle in terms of a dialectical solution; from the standpoint of pure possibility the world can be either finite or infinite. It is very remarkable that this problem should now have cropped up again in a more acute form as a dispute between Leibniz and Newton. The inability of the natural sciences, which have solved so many problems, to deal with this one or even to make a start with it certainly made a deep impression on Kant. We may notice incidentally that the problem is not exhausted in these two theories of a finite and infinite world. Both theories presuppose an unchangeable infinite time in which the world runs either a finite or an infinite course. But there is a third possibility. It might be the case that the world has only existed for a finite number of years but that time itself began with the beginning of the world. Plato seems to have considered such a solution in the *Timaeus*. There would then not be a finite world in infinite time, but time would be finite itself. Classical physics had completely disregarded this idea of finite time and also the corresponding idea of finite space. Kant's thought is limited in the same way, and he only considers a finite or an infinite course of the world in infinite time.

The second Antinomy consists of the thesis, 'Every composite substance in the world is made up of simple parts and there exists nothing at all except the simple or what is composed of the simple,' and of the antithesis, 'No composite thing in the world is made up of simple parts and there exists nothing at all in the world which is simple' (A 434, B 462). In this antinomy two trains of thought are joined in a way which is somewhat artificial even though the conjunction had been made long before Kant's time. It turns on the one hand on the atomic theory and the contrary thesis that space is continuously filled with matter; and on the other hand on an attempt to prove the imperishability of the soul. A simple thing, so it is argued in this proof, cannot perish. If therefore it can be successfully proved that the soul is simple, its imperishability has also been proved. The interlinking of these two problems gives to the second Antinomy a very artificial structure. We propose to take the two problems apart again and only to consider the first one, namely the question of the structure of matter.

On this question, as on those of the limitation of the world in space and in time, there have always been two opposed opinions. The discussion is opened in Greek atomic theory by Democritus and Leucippus. According to Greek atomic theory, matter consists of ultimate indivisible particles which are different from each other in shape and which can take up different positions. All differences in material bodies rest on differences in the form and position of atoms, and all chemical changes rest on changes in the position of atoms. Plato's *Timaeus* can also be regarded as an atomic theory. Plato goes further by attempting a deduction, in which the atoms of the five elements are regarded as instances of the five regular solids (*Timaeus*, 53d). In this way both the number of the elements and the mutual interchangeability of the elements could be made theoretically intelligible, although the attempted theory is not in fact completely satisfactory. Aristotle repudiates this atomic theory of matter and teaches the continuous distribution of matter. There are four sublunary elements, earth, water, air, and fire. The number of the elements is deduced from the primary qualities and the four elements are said to be changeable into one another, but no theoretical explanation of these processes is attempted. The fifth element, ether, does not take part in this cycle of changes.

This conflict between the atomic theory and the Aristotelian theory has always been regarded as a fundamental problem. Kant's actual starting-point was probably again the correspondence between Leibniz and Clarke in the first instance, although I do not underestimate in any way the importance of the historical accounts of the problem in Brucker's history of philosophy [3] and Bayle's dictionary (art. Zeno), and the systematic account of it in Wolff. The problem plays a momentous part in the controversy between Leibniz and Newton. Newton stands for a strictly atomic theory and regards matter as composed of indivisible particles which can change their relative positions in empty space. Leibniz upholds the contrary theory of the continuous occupation of space by matter; for him there are neither indivisible bodies nor empty space, and every position in space is filled with infinitely composite matter. It is remarkable that this problem of the structure of matter could not be solved by purely philosophical means, but it was certainly also an exciting insight for Kant that natural science founded on mathematics was also unable to bring about a solution to the problem. The Antinomies are a systematic formulation of this long historical development, in which the particular reference to the controversy between Leibniz and Newton should not be overlooked.

The systematic proofs of each proposition in the Antinomies aim at showing that there is a contradiction in the contrary proposition. In this they give a true picture of the historical situation and in particular of the controversy between Leibniz and Newton, since the arguments of Leibniz and Newton rest in the main on showing that there is a contradiction in the opponent's position.

The antithesis in the first Antinomy states that the world cannot have a beginning in time. To prove this it therefore has to be shown that there is a contradiction in the thesis that the world has had such a beginning. A beginning presupposes that the world runs its course in a time which, until then, was empty. 'For let us assume that it has a beginning. Since a beginning is an existence preceded by a time in which the thing is not, there must have been a preceding time in which the world was not, i.e. an empty time' (A 427, B 455). In this empty time before the beginning of the world there was the passage of time, but no events. But as all parts of time are without distinction similar, all time before the beginning of the world passed exactly similarly, all moments are exactly alike, and there is nothing to distinguish any particular moment as being the one at which the world was to begin. There is thus no sufficient reason why the world should begin at this moment rather than at any other, and the world therefore cannot begin at all if it has to begin at a finite time.

> Now no coming to be of a thing is possible in an empty time, because no part of such a time possesses, in preference to any other, any distinguishing condition of existence in preference to one of non-existence (A 427, B 455).

This is Leibniz's argument in the controversy with Clarke:

One cannot say, as the author does here, that the wisdom of God had good reasons to create this world at such and such a particular time: that particular time considered without the things being an impossible fiction and good reasons for a choice being not to be found, where everything is indiscernible (5th Letter, § 58).

Kant argues in the same way against a limit to the world in space. If the world were finite, it would have to have a relation to infinite space. But such a relation is nothing because it cannot be given (A 429, B 457). Kant alludes explicitly to the nonsensical consequences of such a representation of a finite world in an infinite space.

If we attempt to set one of these two factors outside the other (space outside all appearances), there arise all sorts of empty determinations of outer intuition, which yet are not possible perceptions, e.g. motion or rest of the world in infinite empty space (A 429*n*., B 457*n*.).

Throughout this argument also, Leibniz's principle of sufficient reason is used with effect. The three points are that there is no sufficient reason which could determine the size of the world and in particular its relation to infinite space; that there is no sufficient reason why the world should be at a determinate place in infinite space at a determinate point in time, for all points in space are similar; and finally that the world can have no motion, because there is no sufficient reason why it should move in a determinate direction and with a determinate speed, for in infinite space all directions and all speeds are without distinction similar.

The same three points are made insistently by Leibniz himself. The first, the impossibility of the relation of a finite world to infinite space, he expresses in terms of the relation between the two parts of space, the part which is occupied by matter (the plenum) and the unoccupied part (the vacuum). 'Here is the other argument, grounded upon the necessity of a sufficient reason. It is impossible there should be any principle to determine the proportion of matter or plenum to vacuum, or of vacuum to plenum' (4th Letter, PS.). Similarly, the argument of the antithesis that a finite world must lead to a completely meaningless motion of this finite world in indiscernible space is wholly grounded on Leibniz's argument.

To say that God can cause the whole universe to move forward in a right line, or in any other line, without making otherwise any alteration in it, is another chimerical supposition. For two states indiscernible from each other are the same state; and consequently, it is a change without any change.

> Besides, there is neither rhyme nor reason in it. But God does nothing without reason; and it is impossible there should be any here (4th Letter, §13).

And finally, Leibniz shows in the same way that a finite world must be somewhere in infinite space but that there can be no reason for it to be at any determinate place.

> And I prove it thus. Space is something absolutely uniform throughout; and without the things placed in it, one point of space does not absolutely differ from another point of space. Now from hence it follows (supposing space to be something in itself, besides the order of bodies among themselves) that it is impossible there should be a reason, why God . . . should have placed bodies in space thus and not otherwise (3rd Letter, § 5).

The thesis of the first Antinomy states that the world has a beginning in time and in space, and it therefore has to be shown that the contrary thesis of a temporally and spatially infinite world is self-contradictory. Now for those who uphold the thesis—what we saw of Aquinas is also true of Newton and Clarke—it is generally the case that the finiteness of the world is accepted as a fact, and that no a priori proof of its finiteness is ever attempted. But in spite of this, the way of proving the thesis that we are considering here not only takes up an old philosophical point, but also goes further and reveals, if I am right, the actual foundation of the thesis both for Aquinas and for Newton. This method of proof is grounded on the impossibility of an actual infinite. There is no actual infinite, 'an infinite past world-series is therefore impossible' (A 426, B 454).

The impossibility of an actual infinite rests in the last resort on the world being created by God and, as God-created, being a world that is finite throughout.

We can see therefore in the first Antinomy how the Kantian proofs rest on elevating the trains of thought that were historically present into a pure systematic form. If we were to consider the other three Antinomies from the same point of view we should be able to continue to bring out the same connection between the systematic and the historical aspects.

§ 8

The Problem of Antinomies in Theory of Science

The Kantian doctrine of antinomies has been very differently judged at different times. Hegel, for instance, was strongly critical, and in Neo-Kantianism the arguments of the Transcendental Analytic, which tend to a systematic conclusion, were given preference over the exciting but problematical arguments of the Transcendental Dialectic. This judgment naturally depends on how the problem of antinomies itself is assessed. Anyone who is inclined to recognise the problem of antinomies as a genuine systematic problem and perhaps even to allow it fundamental importance, will be pre-disposed to give great weight to what Kant has to say about it. Anyone who rejects the problem of antinomies, or is at least inclined to regard it as a problem of secondary importance, will first examine the formal correctness of the Kantian arguments and will certainly find things in them to object to. It is therefore understandable that with the re-appearance of the problem of antinomies in mathematics and in physics this aspect of Kantian thought should have been given fresh attention, and an informed account of the structure of the problem of antinomies, as we know it today, is perhaps more important than a formal examination of the particular arguments. I should like to think that this is true of every dialectic, whether of Zeno, Plato, Kant, or Hegel. The place of the problem of antinomies in philosophy is so difficult to determine that any account of it must necessarily be unsatisfactory. Every interpretation should therefore try to bring out not the formal correctness of what is stated but rather the genuine ground of the problems.

From the point of view of present-day fundamental theory of science, the problem as a whole, as we see it, turns on two areas in which antinomies originate, and both of these underlie the Kantian doctrine of antinomies. The one source of antinomies lies in the possibility of presenting the same facts in different ways and even in some circumstances in mutually contradictory ways, and the other in the concept of totality.

The question of different presentations concerns both the antinomy of finiteness and infinity and the antinomy of atomic and continuous matter.

Whether a given region is finite or infinite seems in many cases to depend on the form of presentation. If one raises the question whether there is a finite lower limit of temperature or whether temperature continues downwards to infinity, the answer seems at first to be given without further question. Temperature has unambiguously a lower limit, namely the absolute zero at − 273·1 degrees. On this there seems to be no possibility of doubt. A lower temperature than this cannot be reached, and even the absolute zero itself cannot be reached. Here we certainly have an indisputable experimental result, but the question is whether this result can be expressed in a different way; and in fact by a simple, purely mathematical transformation a result that is formally quite different can be reached. If normal temperature is transformed into absolute temperature by means of the equation $T = t + 273{\cdot}1$, it is then possible to proceed to the further transformation $T' = \log_e T$. The whole transformation is now $T' = \log_e(t + 273{\cdot}1)$. As these are purely mathematical transformations which cannot affect the experimental results, I can regard either t or T or T' as the 'real temperature'. For many purposes T' would be much more suitable; for instance, the impossibility of reaching absolute zero is represented much better by T'.[1] For our present considerations this possibility of a purely mathematical transformation is very important. For while t and T have a finite lower limit, T' has no finite lower limit. This means that whether temperature is finite or infinite depends on an arbitrary mathematical transformation and cannot therefore be an absolute property of temperature itself. It is we ourselves who are able to represent the lower limit of temperature as either finite or infinite.

Now the same is probably true of any experimental result, namely that it can be presented in different forms. We are accustomed to reproducing every experiment as a curve, without further consideration, but one must bear in mind that usually the experimental data themselves do not represent a uni-dimensional curve but a band of a certain breadth. There is certainly good reason for reducing such a band to a curve, but any mathematical presentation which reproduces the band is a possible presentation of the experiment. Here, naturally, a great number of different presentations are possible, in particular presentation by continuous or discontinuous functions. If this is possible it is obvious that the continuity or discontinuity of the process that is being presented cannot be inferred from the continuity or discontinuity of the functions by which it is presented. In some circumstances therefore continuity or discontinuity belong to the means of presentation; it is we who have the choice of presenting the process in a continuous or discontinuous way. Continuity or discontinuity cannot be regarded as properties of the process, or at least the properties of the processes themselves cannot be inferred from the properties of the functions by which they are presented.

The connection with the second Antinomy is plain, and it appears again even more clearly in the following particular problem. Newtonian mechanics can be presented in two different forms. One form yields the mechanics of systems of points, the other form the mechanics of continua. There has been much controversy about the respects in which these two forms are identical and the respects in which they differ, and in particular about whether one of the forms is the fundamental one and if so, which. This conflict has not reached any decision and, if we are right, cannot reach any. The fact is that Newtonian mechanics is independent of the underlying hypothesis about the structure of matter. One must of course take a determinate hypothesis as the underlying one. If the atomic theory is taken as the underlying hypothesis, one gets the mechanics of systems of points; if the continuous distribution of matter is taken as the underlying hypothesis, one gets the mechanics of continua.[2] The Newtonian propositions themselves are invariant in respect of this variation. This means that classical physics contains no means of deciding between the atomic theory of matter and the continuum theory of matter, and the controversy over this point between Newton and Leibniz is only the historical expression of this. On the basis of classical mechanics one can choose freely between these hypotheses, and matter can be presented as atomic or as continuous. Atomic structure and continuity are not absolute properties, but only manners of presentation.

The subjects that we have so far been discussing were deliberately chosen so as to be verifiable in classical physics. The problem has acquired much greater importance in quantum physics, where the Kantian doctrine of antinomies has excited the interest of present day physicists. C. F. von Weizsäcker's article on the relation of quantum physics to Kantian philosophy may be specially mentioned.[3] The dual nature of the objects of quantum physics seems to bring out the problem of antinomies much more clearly. The significance of the problem for physics itself also seems to go much deeper. In classical physics the point is that the same experimental result can be presented in different ways. In quantum physics, to say that light, for instance, has a dual nature means not merely that there are different possibilities of presenting identical experimental findings but that the dual nature of light affects the experimental result. Whether the light appears as wave or as corpuscle depends on how the experiment is conducted. We are convinced that there are problems here which have a fundamental connection with the Kantian doctrine of antinomies, but that there will have to be long and tedious work by physicists and philosophers before these problems can be really cleared up. For our interpretation it seems to be desirable to limit ourselves to the more easily intelligible problems of classical physics. The problem of antinomies in classical physics is restricted, as we saw, to the means of presentation, but we have the benefit of general agreement about the subject, and it is certain that for a whole series of properties it is not permissible to make an inference from a property of the means of presentation to a corresponding property in the processes. Finiteness and infinity, atomic structure and continuous occupation of space, are in the first instance structures of the means of presentation and they may not be naïvely transferred to the processes themselves.

The second region of the problems of antinomies is the antinomies of modern mathematics. Kant's antinomies belong to physics or to metaphysics and Kant himself would have been very much disturbed by the idea that there might be antinomies in mathematics. The discovery of mathematical antinomies in the theory of sets did in fact justifiably create a sensation. They became even more important when a presentation of the whole of the fundamental propositions of mathematics was attempted by the methods of mathematical logic. All these endeavours are combined in *Principia Mathematica* and it is therefore understandable that Bertrand Russell, one of the two authors of this work, should have made a decisive contribution to the problem of antinomies. The connection with the Kantian doctrine of antinomies has often been noticed and I myself am greatly indebted to Ernst Zermelo for drawing attention to it in his lectures.[4]

The problem of mathematical antinomies has now been worked out to the point that it is possible to set up antinomies according to rule. We choose for our consideration the class of all classes, which can be shown to contain an antinomy. If one imagines to oneself that there is a class of all classes M = (A, B, C . . .), this class of all classes is not the class of all classes since I can at once give a class which cannot be contained in this class of all classes. In order to do this I only need to extract from M any element, e.g. A, and to make a new class (M,A) out of this element and the class M itself. This class (M,A) certainly cannot be contained in M since it is only formed on the basis of the formation of M. The class M is thus certainly not the class of all classes since it does not contain the class (M,A). If the class (M,A) is contained in the class M then it can easily be shown that the class M itself is contained in M, but if I allow that a class can be contained in itself the contradiction follows. In *Principia Mathematica* Russell expressed this antinomy in its most elegant form.[5]

It will be seen that the two sides of this antinomy are very intimately connected. Both rest on considering the class of all classes, which has the particular classes as its elements, as itself an element in its turn in the new formation of a class. The one side of the antinomy rests on this, since M is regarded as an element in the new class (M,A), and the other side of the antinomy rests on this since a class is regarded in general as an element of itself. We are thus concerned with the concept of totality. I consider all classes and form the concept of their totality, namely the class of all classes. But then I regard this totality as an element in the formation of a new class. A concept of a totality is again formed and this new totality is again used as an element. This conflict between concluding and beginning anew, between forming a totality and using this totality as a new element, is the actual ground of the antinomy. It is this conflict that gives the connection with the Kantian antinomies. Kant saw quite clearly that the antinomies rest on this antithesis between making a conclusion and going beyond the conclusion. In principle this had already been seen by Archytas, when he wanted to go to the end of the world and then stretch out his arm, and the antinomies of Zeno rest on the same basis.

There can be no doubt that a solution in mathematics must be sought. Three main solutions have been proposed, the intuitionist theory of mathematics of Brouwer, Zermelo's theory, and Russell's theory of types. All the solutions are intended to make this conflict impossible. They forbid the re-using of a determinate totality as an element. The antinomies can be cancelled by such a prohibition, but fundamental parts of mathematics or at least fundamental concepts have to be cancelled at the same time. It appears from the fact that none of the three solutions has established itself that none of them can be regarded as finally satisfactory.

Kant would doubtless have raised the very greatest objection to admitting antinomies in mathematics. But there can be no doubt that such antinomies occur and they reinforce the Kantian doctrine of antinomies since they show that at least certain totalities contain antinomies.

§ 9

The Transcendental Ideality of the World

We have followed the antinomies in their historical development and discussed them from the point of view of the problems of modern theory of science. It remains to describe the ontological consequences of the Kantian doctrine of antinomies. The ontological result cannot be more succinctly stated than by Kant himself.

> From this antinomy we can however obtain a true advantage, not indeed a dogmatic but a critical and doctrinal one. It affords indirect proof of the transcendental ideality of appearances, if there should be anyone who is not satisfied with the direct proof given in the Transcendental Aesthetic. This proof would consist in the following dilemma. If the world is a whole existing in itself, it is either finite or infinite. But both alternatives are false (as shown in the proofs of the antithesis and the thesis respectively). It is therefore false that the world (the totality of all appearances) is a whole existing in itself. From this it then follows that appearances in general are nothing but our representations—which is just what is meant by their transcendental ideality (A 506, B 534).

The antinomies constitute an independent proof of the Kantian thesis of the ideality of appearances. Hence they played an important part in the development of critical idealism, and in its later defence they play a fundamental part. In the *Prolegomena*, for instance, Kant proposes that the *Critique* shall be examined on the basis of the antinomies alone.

> Now he has the freedom to choose one of these eight propositions [the theses and antitheses of the Antinomies] at his pleasure, to accept it without proof . . . and then attack my proof of the contrary. If I can still rescue this proposition and show in this way that according to principles which every dogmatic metaphysics must necessarily recognise, the contrary of the proposition he has adopted can be proved equally clearly, then it is established that there is a hereditary fault in metaphysics which cannot be explained, much less removed, except by ascending to its birthplace, pure reason itself (IV,379).

The antinomies are in fact, as is claimed here in the *Prolegomena*, a proof of transcendental idealism.

In order to follow the Kantian trains of thought in their original order, let us consider first the definition of the world. In the passage from the Transcendental Dialectic quoted above, the world is defined as the totality of all appearances. 'Appearance' has a simple meaning and an ontological meaning for Kant. In the simple meaning, 'appearance' delimits the sphere of objects which appear, and therefore denotes nature as it is given in ordinary simple experience. We speak for instance of the appearance of a comet. But the expression 'the appearance of a comet' has an overtone of the unusual or the wonderful. If this overtone is removed, we can say that every eclipse of the sun or the moon is an appearance, the sun appears every morning anew, sunrise is an appearance, so is sunset, and so are the whole course of the sun, the moon, and the stars, and the course of nature as a whole.

All these are appearances and can now be regarded from three points of view. They are first characterised by occurring in space and time, all of them in time and nearly all of them in space. In so far as appearances are regarded from this point of view, space and time are seen to be forms of appearances. But appearances can also be considered from the point of view of whether and how they are connected together. Appearances are then seen to be connected in an uninterrupted causal chain and in an uninterrupted chain of effects, and thus all appearances are in uninterrupted connection with each other according to laws. When Kant considers appearances from the point of view of this connection according to laws, he calls them, as we shall explain at length in the next chapter, a nature. Nature is the connection of appearances according to laws. Thus space and time are the forms of appearances, whereas nature expresses the inner connection of appearances, the fact that every appearance is connected with every other appearance. Finally, appearances can be considered from the point of view of whether they form a whole, a totality, and when Kant is considering appearances from the point of view of totality, he calls them a world. 'World' thus means the totality of appearances, but essentially with reference to totality as such. The first main question of the *Critique* was, what are the forms of appearances, what are space and time; the second main question was, what is the connection of appearances according to laws, what is nature; and finally the third question is, what is the whole, the totality of appearances, what is the world? This is expounded systematically by Kant, as we saw, but it emerges particularly vividly from for instance the first Antinomy. For if I ask about the first beginning or the ultimate boundary of the world, then my question about the world is about its wholeness as such.

Transcendental cosmology thus poses the question, 'what is the world?', and there are corresponding questions to this one in transcendental psychology and transcendental theology, namely 'what is the soul?', 'what is God?'

If one tries to approach the question 'what is the world?' through language, the substantive 'the world' seems to correspond to other substantives such as 'the sun', 'the moon', 'the earth'. The words seem to suggest that the world is the same sort of thing as the sun, the moon, or the earth, and that is what is in fact intended. For the ordinary simple conception of the world that has precipitated itself in language, there is in fact only one mode of existence. A man, an animal, a plant, a house, a ship, are all considered as one mode of existence, and all existence is determined on this basis, not only the sun and the moon and the earth but also the world and space and the soul and God. In Greek mythological thinking there is Zeus and Poseidon just as there is Agamemnon and Achilles; certainly they are different but at bottom they have the same existence. At bottom, the world is taken as existing in the same way as the sun, the moon, and the earth. Here we meet the same difficulty as with space. In the case of space, if it was assumed to be a substance, then something that denoted only the totality of all spatial things was posited as a further spatial thing alongside the others. In the same way the world, which is only the whole, the totality of everything in the world, is posited as a further thing alongside the other things. This attitude continues long into educated Greek thinking, in which the cosmos is described as a living being or as a sphere.

This picture is destroyed in the new natural science grounded on mechanics, but what now becomes of the world? It is no longer a determinate tangible existent, but is the whole of this course of events according to law. But what is this whole? This is the question which Kant poses in the Transcendental Dialectic. If the world itself were a thing, it would have to have its share in the disjunctive properties of which either the one or its opposite must be attributed to everything, and the world itself would have to be either finite or infinite. But since the antinomies have shown that it is either both at once or neither, the world cannot have the same existence as things to which one of every pair of disjunctive properties must be attributed. The progress of philosophical thought takes place in the same way in respect of the world as in respect of space. In the case of space, philosophical reflection started from the fact that if space is called a thing, it has to be a spatial thing, and would have to share in the properties of spatial things and would have to be somewhere. Similarly the world, if it is taken to be a thing like things in the world, must itself share in these properties and must be either finite or infinite; and as this is not true of the world, it must have a different status.

The world is not a thing existing in itself but is the totality of things. Now the antinomies show further that a unity, such as this is, can be given in two different ways. The unity of the world can be thought both as a finite and as an infinite whole. But since the world, even considered solely as a unity, cannot be both finite and infinite, this unity must come from us. It is we ourselves who think the world as a unity. This unity occurs in two different forms simply because we can think this kind of unity in two different ways. Finiteness and infinity, atoms and continuous matter, are only our way of thinking the unity of the world, and from the fact that we are able to or have to think of the world as finite or infinite, there is no need at all for it to follow that the world itself is finite or infinite. We therefore interpret that Kantian doctrine of antinomies, on the basis of modern theory of science, to mean that the different concepts finiteness and infinity, atoms and continuous matter, are only the ways in which we think of the world, or to use a more appropriate expression, the ways in which we exhibit the world. From the fact that we can exhibit the world in different ways it follows that these are not objective properties of the world but only our presentations of it.

This duality of finiteness and infinity is made intelligible by Kant through the conflict of understanding and reason, understanding always setting a final boundary and final conclusion, and reason always going beyond such a final conclusion (A 529, B 557). There is room for doubt as to whether Kant's explanation of the antinomies in terms of the difference between understanding and reason as different faculties of knowledge is convincing in all its details; though with the understanding of the mechanism of antinomies that we have gained from theory of classes and the foundations of mathematics, we are bound to see that Kant had put his finger on something genuine. To support this interpretation of the antinomies I can call on the *Metaphysical First Grounds of the Natural Sciences*, where the problem is taken further. There the part of the second Antinomy that we have been considering appears again in a pure form as the question as to the structure of matter. The solution given in the *Metaphysical First Grounds* seems to me to be that in the case of matter the concept of atoms and the concept of the continuum present two equally legitimate possibilities. In so far as the natural sciences confine themselves to one possibility, this was in Kant's time the atomic theory. The task of philosophy, as theory of science, is to draw the attention of scientists to the other possibility and thus to acquaint the natural sciences with the full range of the conceptual possibilities (IV,524). We understand this point of view very much better today, and it therefore seems to us regrettable that Kant does not always entirely keep to the high level of this purely dialectical thesis, though we do not deny that such a point of view is as abstract as it is subtle. Kant lets himself be carried away quite often by his personal inclination towards the concept of infinity and the hypothesis of a continuum and takes up only one of the possible standpoints (in the *Metaphysical First Grounds* and elsewhere), although according to his own best intention he ought to have reached the position of regarding the two antinomically possible standpoints, continuously and without exception, as in every way equally legitimate.

The great importance of the consequences that follow from the doctrine of antinomies is also shown in the problem which might be called the theological foundation of truth. On the basis of the antinomies it seems hardly possible to call truth, and especially truth in the natural sciences, the thoughts of God. It did indeed seem for a long time that to call truth the thoughts of God was something that had been wholly removed from the region of the thinkable, and only the most recent developments in physics have brought such considerations to life again. They are there in an extraordinarily transparent form in the correspondence between Leibniz and Clarke, and it was probably in this form that Kant came into conflict with them. Leibniz and Newton are both in fact almost completely in the position of regarding truth as the thoughts of God. When for instance the question as to whether matter consists of atoms is discussed in the correspondence, Leibniz objects that this is not possible, since the atoms are placed in a completely homogeneous empty space and there is therefore no means of determining what the diameter of the atoms should be and no reason why the atoms should be exactly so big and not all ten times bigger. God has therefore not created atoms, since he does nothing without reason. Atoms were not even contained in a divine plan for a world because even in such a plan, that is to say in the pure thoughts of God, there would be no sufficient reason for the size of the atoms. The world cannot therefore consist of atoms because God could neither think nor create an atomic world.

Newton's objection is fundamentally that on the contrary the fact that the world consists of atoms and is finite in space and time rests on the world being so created by God. In order to answer a purely physical question as to whether matter consists of atoms, Leibniz and Newton go back to the will of God, and this involves going back from God's will to his thinking. The world is as it is because God has so created it according to his divine plan of creation. This approach means that a decision has been taken as to how scientific thought is to be described. If the world is grasped in scientific thought as it really is—for example as consisting of atoms—then this understanding is also an insight into the divine plan of the world. Since God thinks his plan of creation continuously, this insight into God's plan is at the same time a repetition of God's thinking. Leibniz's saying, 'when God calculates and cogitates a world is made',[1] gains the further meaning that the calculations of our mathematicians and physicists are an imitation of God's calculations and thus a repetition of God's thinking. Leibniz knew very well that there was a connection between his point of view and older attitudes and he quite consciously takes up the doctrine of Augustin and Malebranche that we see all things in God (VI,574, on Locke's *Examination of P. Malebranche's opinion of seeing all things in God*).

This view of the nature of mathematical and scientific thought is in fact very old; and Plato stated it incomparably well. For Plato it is God who originally knows and recognises the ideas, and our soul's knowledge of them originates from its eternal and timeless existence. It may expect to return after death to pure intuition of ideas. To the extent that the soul tries on this earth to see ideas, it is approaching its previous condition, indeed approaching divine being, and Plato speaks of the soul becoming like God in its intuition of ideas (*Theaetetus*, 176b, *Laws*, 716d).

This approach is again expressed very clearly in respect of scientific knowledge in the *Timaeus*. The world of ideas is the divine plan according to which this world is created, and as such it contains the ideas of the elements, the ideas of earth, water, air, and fire. A natural science which achieves knowledge of these ideas is ascending to the divine plan of the world.

This view of truth as insight into the divine plan of creation that is worked out by Plato, Augustin, and Leibniz could be called the theological foundation of truth. It is not hard to see that from the Christian point of view this must look in some ways very attractive.

The great influence of the *Timaeus* probably rests in the main on this attitude. It was Augustin more than any other man who followed up the idea of a connection between the truth as we can know it and God's thinking. But authoritative as Augustin's voice may be, weighty objections have to be put forward from a Christian point of view against this identification of human knowledge with divine thinking. Plato really brought out the decisive point. This conception of knowledge means, as we saw, that man can become like God, which is not a tolerable consequence from the Christian point of view. Aquinas placed important limitations on this somewhat naïve approach and it was probably this problem of the relation of human thinking to divine thinking that forced him to seek a different way from that of Plato and Augustin. Luther completely rejects the identification of human thinking with divine thinking.

The significance of the Kantian doctrine of antinomies in this context is that in it a theological foundation for truth is rejected on purely philosophical grounds. If it is true that the world is both finite and infinite, that it contains both atoms and matter which occupies space continuously, the never-ending conflict of these antinomic characters of the world could not come from God's thinking. They must in principle come from human thinking. Kant sees very early that Malebranche's theory that we see all things in God is impossible for a thinking which suffers from a conflict of antinomies. It was of course well known to Kant that Leibniz had taken up Malebranche's doctrine, and he therefore opposes this view explicitly in the scholium to Section IV of the *Dissertation* of 1770 (II,410). The full development of the antinomies in the Transcendental Dialectic makes it completely impossible to hold the old conception of mathematical and scientific truth as an insight into the divine plan of the world.

The Kantian antinomies thus show themselves to be extremely far-reaching. If one can avoid being held up by formal examination of the particular inferences, and can give unhindered attention to the essential problem, it is hardly possible to escape these consequences. Certainly there are many things that could be improved as regards the formal correctness of the arguments, but it must be remembered how difficult these problems are. Kant certainly overestimated the finality of his solution, as for instance when he said that there are four and only four antinomies and that there can be only these four. On the other hand we must keep in mind that we ourselves are still a long way from solving the problems, and indeed it has to remain undecided whether a final solution to the problem of antinomies is possible at all. But this does not alter the fact that, for all its difficulty, the problem of antinomies has always been highly suggestive. This is true of the Eleatic dialectic as it was developed by Zeno and especially of the dialectic that was worked out by Plato in the *Parmenides* and the *Sophist*. The Hegelian dialectic, on the other hand, seems to move away from the goals of the Kantian doctrine of antinomies. This is shown in the fact that Hegel regards the definition of truth as the thoughts of God as completely compatible with the dialectic and regards the task of the dialectic as 'to present God as he is in his eternal essence, before the creation of nature and of finite spirit' (IV,45*f*). In modern fundamental theory, both in the mathematical and in the physical problem, the antinomies have shown their old power again. As regards the physical antinomies, it is hardly possible to escape the consequence that these antinomies force us to regard physical theories as models produced by men. But to fall into pure conventionalism, as some schools of modern fundamental theory have done, is surely to go too far. The physical antinomies certainly show that the theories of natural science are constructed by men and that they only have the character of models. They certainly show that in the construction of scientific concepts a certain part is ascribable to human freedom and that such theories could have been constructed differently. But they do not necessarily prove that the whole of the concepts of natural science are infected with human arbitrariness and that the natural sciences rest entirely on conventions.

The same is true of the mathematical antinomies, one of which we have presented by way of example. They prove that the objects of mathematics cannot be things existing in themselves in a naïve sense. This consequence or one like it has in fact been drawn in universal terms, and it has been seen that statements about the existence of classes can only be made with very great caution. But here too it may be questioned whether the people who, on the basis of the problems of axiomatics and of the mathematical antinomies, have conceived mathematics as mere play, have not gone too far. However great a part the spontaneity of human thought in mathematics may play in this problem, it is not necessary to transform mathematics wholly into an arbitrary game.

We can perhaps summarise the problems of the Kantian doctrine of antinomies, which we have discussed in terms of the first two

Antinomies, in the following way. They turn on certain determinate physical problems, the spatial and temporal extent of the world and the structure of matter. For conceiving these problems we have at our disposal in each case two concepts, finite and infinite space, finite and infinite time, atomic theory and the theory of a continuum. This having at our disposal can be conceived as purely logical; purely logically speaking there are the two concepts and both the opposing concepts are free from contradiction. Or this having at our disposal can be referred to the actual structure of human beings; we are made in such a way that we have such concepts at our disposal and it is not unthinkable that there might be other beings who could have their experience of the boundaries of space and time or the structure of matter in other terms. In this case a temporal change in the equipment of human beings is thinkable. Either man as such or man at a certain time would then be so made as to have at his disposal precisely these pairs of concepts. Finally, this having at our disposal can be interpreted in terms of pure transcendental logic; this would mean that the theoretical systems of physics can be built up both on the one and on the other concept. Whatever may be the solution, the antinomic character of these concepts means that there is here in theory of science an a priori, and in ontology a limitation. If we have at our disposal, in any of these possible senses, only these antinomic concepts, it follows that we have to conceive our problems in terms of one or other of the two concepts; we always have to regard the world either as finite or as infinite and matter always as atoms or as a continuum. But this also implies a limitation in ontology. The antinomic concepts have a priori validity but they cannot be applied to things in themselves. Things obviously cannot themselves be both at once, and it is very improbable that they should happen to fall on one side or the other of the alternative. No other consequence can therefore be drawn but this, namely that both the contradictory concepts are models by means of which we think things. It is not necessary for things to be as these models present them, they cannot be both finite and infinite, and it is very improbable that they should merely happen to be one or the other. It thus follows from the doctrine of antinomies that the concepts in question have the character of models, and from this there follows the transcendental ideality of the world.

CHAPTER III

NATURE

§ 10

Biology, Theology, and Mechanics as the Goals of Science in Ancient, Mediaeval, and Modern Times

THE Transcendental Analytic is the climax of the *Critique of Pure Reason*. One can well understand that the German Idealists' and the Neo-Kantians' interpretations of Kant should have started almost exclusively from the Transcendental Analytic, and Martin Heidegger's Kant is also essentially the Kant of the Analytic. We hope that we have shown the due importance of the other two main parts of the *Critique*, the Transcendental Aesthetic and the Transcendental Dialectic; an interpretation of Kant should try to keep these three groups of problems, the Aesthetic, the Analytic, and the Dialectic, in balance. But in fact all the threads run together in the Transcendental Analytic. Empirical psychology, and transcendental psychology, formal logic and transcendental logic, theory of categories and ontology, all contribute to bringing the development of Kantian thought to its highest point. The Transcendental Analytic not only has its own new field of problems; it is also a summing-up of the whole. It has its own field of problems in that its theme is the being of nature, and it is a summing-up of the whole in that it treats of the being of unity in general. But space and time as well as the cosmological concepts which we have discussed in Chapter II are unities, and in so far as they are unities they are in a certain sense discussed a second time in the Transcendental Analytic. The difficulty here was seen a long time ago. It is not the case that Kant finishes with space and time in the Aesthetic, leaving himself free to turn to new problems in the Analytic. Without reference to the principal themes of the Analytic, unity in general, object in general, subject in general, space and time cannot be understood. Recognising this, Cohen and Natorp considered taking space and time out of their special position in the Transcendental Aesthetic and incorporating them into the Analytic. We propose to try another method. We shall treat the problems of nature, which are in some ways regional, by themselves in this chapter, and extract the problems of unity in general for treatment separately in the following chapter.

We turn first to the problem of nature, and to this end we first have to show that nature has by no means a completely identical meaning for Aristotle, Aquinas and Kant.

Aristotle classifies the sciences as a whole into logic, physics and ethics. In this division mathematics is included under physics, in accordance with the Aristotelian ontology. The nature that is treated in the Aristotelian physics is that which contains its ground of movement in itself, as is stated in Aristotle's definition in the second book (*Physics* II.1.192b13). If this definition is examined, Aristotle's physics is seen to be the science of the living, or to use a modern, admittedly rather narrow term, Aristotelian physics is in respect of its field of objects biology. Of course the term 'living' has a greater extension for the Greeks than it has for us. For the Greeks not only men and animals and plants are alive, but also rivers, springs, rocks, the elements, the stars. We ourselves still have the power of feeling at least some things from this wider field as living things—precious stones, perhaps, and water. Aristotle always thinks of the concept of living things in this wider extension, and if the doctrines of the elements, of motion, of the stars, belong to physics, they belong to it because the elements and motion and the stars are living. We sometimes speak of the early Greek philosophers as being simply purveyors of natural knowledge, investigators of nature. But the philosophy of Aristotle is also in one of its most essential parts natural science, and as natural science it is a science of the living.

In the Middle Ages theology steps into the first place. Not that there is no natural science. Aristotelian natural science is cultivated with great care, and in important respects the new natural science prepared. But in the Scholastics' own thinking nature appears primarily under a theological aspect, as the created works of God, as *creatura*. This conception of nature as *creatura* has no pejorative sense whatever; on the contrary, as the created works of God nature becomes something beautiful and perfect in itself and attains an independent dignity as *imago* and as *vestigium Dei*.

With the new natural science there comes a new way of seeing things, although the great discoverers and investigators of the new natural science keep to the old ways of talking. For Kepler, Galileo, Newton, and Leibniz, nature is still something living and is still the work of God's creation. But even though the great scientists may live their own lives out with the old ideas, natural science has been fundamentally altered and has become mechanics. We can see this in the greatest achievement of the new natural science, the discovery of the laws of the motion of the heavenly bodies. The question of the motion of the heavenly bodies is the primary theme of natural science, both historically and systematically. Through many thousand years the best brains of humanity had looked up enquiringly to the skies and searched for the laws of the motions of the heavenly bodies. Long before the laws governing the motions had been discovered, their regularity as such had been seen. The law governing the motions of the stars, even though as yet undiscovered, gave the basic form for a law of nature, indeed for law in general. The geocentric solution of the Greek astronomers is certainly a solution of genius, but one cannot fail to see that it is in many ways something very artificial. We have to bring to mind again and again how important was the discovery and definitive presentation of the laws of motion, for all time the greatest of scientific discoveries. But how abstract were the means by which the solution was reached! Concepts of living things disappear entirely, no appeal to God's acts is possible, it is in truth a disenchanted world. We are only left with masses moving according to mechanical laws; natural science has become mechanics. However much the great scientists may live by the old ideas (as we have shown at length in the previous chapter) physics has become mechanical natural science, and is to continue irresistibly the way of the disenchantment of the world. We must always keep in mind that when Kant asks about the existence of nature, he is not asking about the existence of the living nor about the existence of creation, but about the existence of nature as it is presented in pure mathematical natural science—in mechanics.

§ 11

Nature in Kant as Newtonian Nature

The concept of nature in the *Critique of Pure Reason* is, as we have seen, limited in that the nature that is discussed there does not include the living; nature for the *Critique* is only the object of mechanics. We now have to show that there is also a further limitation. The Kantian concept of nature does not bear on plants and animals and mountains and not even on the sun, the moon, and the stars, but on conformity to law as such. Kant defines nature as 'the conformity of appearances to law in space and time' (B 165). With this we are given the third mode of appearances. We were given first space and time as forms of appearances, then the world as the whole of appearances, and now nature as the conformity of appearances to law. In accordance with the limitation explained in the previous section Kant is only concerned with mechanical conformity to law, and the nature whose existence he is asking about is thus mechanical conformity to law. This is all that he is asking about. The Kantian proposition which we are trying to understand, namely that the understanding produces nature, does not say that the understanding produces trees or plants or mountains and does not say that the understanding produces the sun or the stars. It merely says that the understanding produces mechanical conformity to law, and that only in respect of its form.

It will help us to further our explanation if we form the concept of a Newtonian nature. By a Newtonian nature we mean one that is completely determined according to Newtonian laws. Such a nature is a purely mechanical system and in the sense of Newtonian mechanics a closed system. By forming this concept we can separate comparatively clearly problems which are usually severely confused and even in the Transcendental Analytic are very much intertwined. We can first ask the categorial question as to the basic concepts which hold for a Newtonian nature or the basic principles which are valid in it. We can ask the further question, whether the nature that we experience is a Newtonian nature; this would correspond to the question, whether space is a Euclidean space. Finally we can ask the properly ontological question as to the mode of existence of this Newtonian nature, and this ontological question corresponds to the ontological question as to the being of space, namely of Euclidean space.

We recognise first that Newtonian nature is independent of the structure of matter, as we have already shown when discussing the problem of antinomies. We can either assume that matter consists of discrete points distributed in space which are endowed with mass, or we can assume that there are atoms properly speaking, bodies of a finite diameter in empty space, or we can assume matter that is continuously distributed. We either get the mechanics of a system of point-masses, or we get the mechanics of continua, and we regard these two forms of mechanics as equivalent. We will now consider the mechanics of a system of point-masses and assume points endowed with mass as a basis for our argument. The simplest model of a Newtonian world is now a single point-mass, a point which is thought of as having a certain mass. The motion of this point through space will continue with a constant velocity and in a constant direction. Although fundamentally unitary, this falls into two fundamental concepts or two fundamental laws, the law of substance and the law of causation. The expression of the law of substance lies in the fact that even in this small world certain factors—in fact nearly all factors—are in principle constant. Not only mass remains constant but also velocity, direction, impulse, and energy. The demands of the Kantian conception of substance are satisfied if any quantity at all remains constant, and it has often been observed that energy can just as well be regarded as this constant, as anything else. We see that even in this, the simplest possible model, mass and velocity and certain products of them, such as impulse and energy, are constant. This also contains the clearest expression of the law of causation. In considering the law of causation one must free oneself, as Kant did, from all anthropomorphic ideas. When considering cause and effect one must not regard a man producing something as the cause and what he produces as the effect. The law of causation means that two temporally different conditions determine each other univocally, so that in every case the previous condition univocally determines the later condition, and the later condition univocally determines the earlier. The law of causation accordingly implies the strict conformity to law of the whole course of events and this conformity to law is given in the most vivid way by the fact that the motion of this single point-mass continues with unchanged direction and velocity through all eternity, so that here every point of time in fact determines every other point in time.

The next simplest model of a Newtonian nature would be a problem involving two bodies, but we proceed to a richer system, namely a planetary system after the fashion of the solar system, that is to say a system with a central body with a mass which is great in relation to the other masses and with for example seven 'planets' rotating in one plane. The laws of conservation are also displayed here. First the mass remains constant. Energy also remains constant, at least if in addition to the concept of kinetic energy the concept of potential energy is introduced, and the sum of the two required to be constant. The pure meaning of the causal law is even clearer if we make use of the two-dimensionality of the system. We can then take time as the third dimension and thus go over to a three dimensional 'space' with two co-ordinates of position and one co-ordinate of time. If we consider, for the time being, the motion of the central point and each of the rotating points one by one, and disregard the motion of the system as a whole, then both the central body and the rotating bodies describe ellipses with centre of gravity as one of the foci. These ellipses are then represented in three dimensional space as helices, whose projections on the plane of the system are ellipses. These curves in three dimensional space have the peculiar property that any adequately defined point and any indefinitely small piece determine univocally the whole course of the curve and thus its condition in every previous and subsequent point of time. Here we have attained a very pure expression of the law of causation—the thoroughgoing regularity in events, which can be reduced to this simple mathematical expression. Not the duality of agent and act and not the duality of cause and effect, but the thoroughgoing determination of events which makes every event, however small, determine all events from all eternity to all eternity—this, properly speaking, is the law of causation. This, the proper sense of causation, can also be grasped intuitively in the fact that in our system, rotation once given will continue indefinitely, which is only the mathematical expression for the fact that the sun and the stars always run the same unchanging courses. Eternal rotation, such as we have here, is the starting-point for the Aristotelian doctrine of the eternal duration of the world which we met in the chapter on the antinomies, and it is intuitively given in a way which no-one can escape.

This then is the proper sense of causality—not the duality of cause and effect but the thoroughgoing determination of the course of events, which in the single body problem leads to endless motion, and in a rotating system to an unchanging repetition of the rotation.

In our more complicated model of the world (first in the two body system) the third fundamental law can be seen, namely the law of reciprocal action. The rotation of the earth round the sun for instance is not merely an effect of the sun on the earth but also an effect of the earth on the sun. Both bodies in the rotary motion are in thoroughgoing reciprocity. This is even clearer in the case of the more complicated model of a planetary system with seven planets. The course of a rotating point is only an ellipse if it is considered solely in relation to the sun. In reality every body is acting on every other body. Not only the central body but also the six other rotating point-masses are acting on the revolving point-mass. We therefore get a very complicated motion which cannot be expressed in a definitive form with the mathematical means at our disposal. With the relations of the masses in the solar system, the reciprocal effects of the other planets are so small that for most purposes they can be disregarded. But for more accurate calculations they have to be taken into account.

We now have in our model the three fundamental laws, the law of substance, the law of causation, and the law of reciprocity. These three fundamental laws still stand if we go over to the universal case and admit in nature a large number of points with very different masses and constants of motion. The law of substance stands, both in respect of mass and in respect of energy. The law of causation stands, in that one state of a purely Newtonian nature at a given point in time determines univocally its whole state at every other previous and subsequent time. A sufficiently informed mind, such as Laplace supposed, could calculate the whole course of the world from all eternity to all eternity, though we have to raise the question whether the world is in fact a Newtonian nature. And finally the law of reciprocity stands. Every point-mass in this pure Newtonian nature affects every other point. In order to calculate the course of any particular point every other point would have to be taken into consideration and Laplace's demon would indeed have to have an unusual gift for mathematics. This is the world which was contemplated with such enthusiasm by those who held the purely mechanical conception of the world, and these are the laws from which it was believed to be possible to draw such limitless metaphysical consequences. The struggle to grasp the significance of this purely Newtonian nature is also one of the essential problems to which the *Critique of Pure Reason* is addressed.

§ 12

The *a priori* Character of Newtonian Nature

Of the three fundamental principles that are valid in Newtonian nature, the principle of substantiality is almost certainly valid outside classical physics, in respect of the conservation of energy. The same is very probably true of the principle of reciprocity, but the principle of causality in the sense in which we have described it is only valid for classical physics and is indeed the peculiar characteristic of classical physics.

Leibniz had recognised this when he proclaimed the principle of sufficient reason as the truly fundamental principle of natural science. We can now see with complete clarity that Leibniz had thereby located the peculiar characteristic of Newtonian physics with the sureness of genius. The sureness of this first approach is astonishing, and yet is only a further example to support a claim for which there is other evidence, that in the really great problems the fundamental positions were almost always reached at the first grasp.

Leibniz talks about the principle of sufficient reason in many places, the clearest of which is probably that in the 2nd Letter to Clarke.

> The great foundation of mathematics is the principle of contradiction, or identity. . . . This single principle is sufficient to demonstrate every part of arithmetic and geometry, that is, all mathematical principles. But in order to proceed from mathematics to natural philosophy, another principle is requisite, as I have observed in my *Theodicaea*: I mean, the principle of a sufficient reason, viz. that nothing happens without a reason why it should be so, rather than otherwise. . . . Now, by that single principle, viz. that there must be a sufficient reason why things should be so and not otherwise, one may demonstrate the divinity and all the other parts of metaphysics or of natural theology; and even in some measure, those principles of natural philosophy, that are independent upon mathematics: I mean the dynamic principles, or the principles of force (2nd Letter, § 1).

The two other principles are also fundamental for Leibniz. The *Monadology* is based from the first on the concept of substance, and the laws of conservation in so far as they are applied not to matter but to energy also play a great part in Leibniz's thought. The law of reciprocity is inseparably woven into the fabric of the *Monadology* and is one of its fundamental principles. Every monad is a reflection of the whole universe and thus also of every other monad. This involves in particular that the events of every monad are reflected in every other monad (§ 56). But important as are the two other fundamental laws, the one which Leibniz worked out most intensively was the principle of sufficient reason, and it includes what is properly speaking the law of causality as its most important particular case.

It is therefore understandable that a lively discussion about the principle of sufficient reason began at once among philosophers, logicians, and natural scientists, among Leibniz's friends as well as among his opponents. Leibniz takes it to be a fundamental principle. But with what right? The question as to its demonstrability soon arose. Christian Wolff gives a proof of it in his *Ontology*, the basic idea of which is taken over into Baumgarten's *Metaphysics*. Others, like Crusius, are more cautious and are doubtful about its demonstrability.[1]

Kant in his early writings studies this problem intensively, and in the *Principiorum primorum cognitionis metaphysicae nova dilucidatio* of 1755 he gives a criticism of a Wolff's proof. Kant shows that in essentials he is keeping to Leibniz's structure. Two great principles, the principle of identity and the principle of sufficient reason, are the foundation of metaphysics. Kant defends the principle of sufficient reason as such against Crusius but rejects the Wolff-Baumgarten proof and shows by a detailed analysis that the principle of sufficient reason cannot be proved from the principle of contradiction (I,396–8). Kant himself gives a new proof, but it falls under the condemnation which he later pronounced over his pre-critical writings (XII,208). The question of the legitimacy of the principle of sufficient reason remains one of Kant's central problems.

The proof of the Second Analogy in the *Critique of Pure Reason* can be regarded as proving the principle of sufficient reason by means of transcendental philosophy. The Second Analogy, the principle of succession in time according to the law of causality, says 'All changes happen according to the law of the connection of cause and effect' (B 232). In the proof itself, Kant roundly calls the Second Analogy the principle of sufficient reason. 'The principle of sufficient reason is thus the ground of possible experience' (A 200, B 246). In the final paragraph of the Analogies Kant returns to the problem as a whole. He notes with unmistakable pride that all the previous attempts at a proof have been unsuccessful.

> For lack of this method, and owing to the erroneous assumption that synthetic propositions, which the empirical employment of the understanding recommends as being its principles, may be proved dogmatically, the attempt has, time and again, been made, though always vainly, to obtain a proof of the principle of sufficient reason (A 217, B 264).

Kant sets out his fundamental attitude to the problem and finally raises the question how propositions of the type in question can be proved at all.

> Had we been trying to prove these analogies dogmatically, that is to say, to prove out of concepts . . . that every event presupposes something in the previous state on which it follows according to a rule . . . all our efforts would have been wholly in vain (A 216, B 263).

A proof is only possible as a transcendental proof, namely as a proof from the possibility of experience. Finally Kant points out that the causal law has always been discussed without reference to the other two laws.

And since the guiding-thread of the categories, which alone can reveal and make noticeable every gap in the understanding, alike in regard to concepts and to principles, has hitherto been lacking, no one has so much as thought of the other two analogies, although use has always tacitly been made of them (A 218, B 265).

In so far as this observation refers to philosophical discussion since Leibniz, it is certainly correct. The causal law is discussed almost alone. If Kant was also thinking of Leibniz himself, the statement could hardly be called well grounded if one considers in Leibniz not merely his actual words but also the substance of what he says. It may be that this is a case in which Kant's pride in his discovery of the table of categories, which for other reasons is not always wholly understandable, has made him make an unjust appreciation.

Historical and systematic considerations are combined in a striking way when Kant returns to the same problem in 1790. Eberhard, professor of philosophy in Halle and a declared follower of the Leibnizian philosophy, had applied to the *Critique of Pure Reason* the old tag, that what was new in it was not good and what was good in it was not new, but was all there already in Leibniz, so that Leibniz's philosophy contained everything that is true in the *Critique*, 'and in addition to that a well-grounded extension of the area of the understanding', as Kant quotes him as saying in his attack (VIII,187). In the essay *On a Discovery, according to which any new Critique of pure Reason has been made superfluous by an older one*, which we shall call the polemic against Eberhard, Kant takes up this attack.

Eberhard tries to prove the principle of sufficient reason in Wolff's sense. Kant gives this proof verbatim and then analyses it in full. He shows that the new proof consists of two parts. The first part is the old Wolff-Baumgarten proof, which he regards as finally refuted. A new argument then follows in the second part of Eberhard's proof. Kant shows in a detailed analysis that this new argument also fails to reach its goal (VIII,195). Any attempt at such a proof is mistaken in principle. For Leibniz there are two first principles, the principle of contradiction (equivalent to the principle of identity) and the principle of sufficient reason. If an attempt is made to prove the principle of sufficient reason there is obviously only the principle of contradiction from which to prove it. But Leibniz's argument would then lose its significance. We saw that Leibniz takes the principle of sufficient reason as an additional principle because the propositions of metaphysics can only be proved on the basis of this further principle, while the propositions of mathematics can be proved from the principle of contradiction alone. But if the principle of sufficient reason can be proved from the principle of contradiction, this distinction collapses and all true propositions can be proved from the principle of contradiction alone (VIII,247).

At the end of the essay Kant returns a second time to the fact that Leibniz introduces the principle of sufficient reason as an additional new principle. Kant says that three peculiarities are characteristic of Leibniz's metaphysics, the first of which is the principle of sufficient reason. But what did Leibniz mean by bringing the principle of sufficient reason into such prominence? The principle itself 'is so universally known and under suitable limitations so obviously clear that the stupidest person could not believe that he had discovered it for the first time; and indeed it has had scorn poured on it for this reason by opponents who have misunderstood it' (VIII,247). On the contrary, the bringing out of this new principle must signify a problem of method. 'For what does it mean to say that there must be further principles in addition to the principle of contradiction?' Here the emphasis is placed on the fact that a further principle is added to the principle of contradiction, and on this fact alone. 'Thus . . . a special principle, different from the principle of contradiction, must be found.' This means that the totality of true propositions does not consist solely of the principle of contradiction and the propositions which can be deduced from it; there are also therefore in Kantian terminology synthetic propositions. 'As the latter kind of propositions are (now at least) called synthetic propositions, Leibniz meant nothing more than this: that in addition to the principle of contradiction (the principle of analytic judgments) there must be a further principle, namely that of synthetic judgments' (VIII,248). Leibniz's distinction between the principle of contradiction and the principle of sufficient reason is thus interpreted by Kant to mean that Leibniz had grasped the difference between analytic and synthetic propositions for the first time, and inversely the Kantian doctrine of the synthetic character of the principles of pure nature is interpreted as the genuine completion of the beginning made by Leibniz. This is certainly a very bold interpretation. It is very rare in the history of philosophy for a great thinker to establish so close a connection with a predecessor, and this bold interpretation contrasts favourably with the many places in the *Critique of Pure Reason* in which Leibniz is criticised, often with little success. I believe all the same that the connection that is being established here is a well-grounded one. The Transcendental Analytic must be understood to mean that not all principles can be deduced from the principle of contradiction and that an original plurality of undeducible principles must be presupposed; and at least in the passage quoted from the correspondence with Clarke, Leibniz was obviously deliberately demanding a plurality of principles.

We can thus regard the Transcendental Analytic, in so far as it is an ontology of nature, as a proof of the causal law. Here the causal law only stands as an example, if the most important one, of the totality of the principles of nature, and 'proof' is taken here in a very broad sense to include factors both of categorial analysis and of ontology proper. The categorial factor signifies the extraction of the principles themselves and the demonstration of their a priori character; and the ontological factor, from the Kantian point of view, signifies the limitation of these principles to appearances and thus the phenomenality of nature. In addition there is the bringing out of the relational character of nature which is very important. A priori character, relational character, and phenomenality or transcendental ideality, these are the goals of the Transcendental Analytic in respect of nature.

The argument of the Transcendental Analytic is carried forward by means of a fourfold table. Kant constructs one after the other the table of the twelve judgments, the table of the twelve categories, the table of the eight (not twelve) schemata, and the table of the eight (not twelve) principles.

We turn to the last table, the Table of Principles. Kant starts from a fourfold schema (A 161, B 200):

1
Axioms
of Intuition

2
Anticipations
of Perception

3
Analogies
of Experience

4
Postulates
of Empirical Thought in general

The Analogies and the Postulates are each divided into three propositions, and to keep to the pattern of the first two tables the Axioms and Anticipations should also have been divided in the same way. There would then have been twelve principles to correspond to the twelve members of each of the first two tables. Fortunately the Kantian system did not quite work out here for once, and there is only one Axiom and one Anticipation, giving eight principles in all. The problems of quantity and quality are discussed first. The principle of the Axioms of Intuition is 'All intuitions are extensive magnitudes' (B 202). The principle of the Anticipations of Perception is 'In all appearances the real, which is an object of sensation, is an intensive quantity' (B 207). The three Analogies issue from the principle of the possibility of experience. The first is the principle of the permanence of substance. 'In all change of appearances substance is permanent, and its quantum in nature is neither increased nor diminished' (B 225). The second Analogy gives the principle of succession in time according to the law of causality. 'All changes happen according to the law of the connection of cause and effect' (B 232). The third Analogy gives the principle of co-existence according to the law of reciprocity or community: 'All substances, in so far as they can be perceived as co-existing in space, are in thoroughgoing reciprocity' (B 256). The table of principles concludes with the three Postulates of Empirical Thought in general:

> 1. That which agrees with the formal conditions of experience (according to intuitions and concepts) is possible. 2. That which is bound up with the material conditions of experience (or sensation) is actual. 3. That which in its connection with the actual is determined according to universal conditions of experience, is (exists as) necessary (B 265).

We follow a suggestion of Cohen's and give our attention first not so much to the particular tables as to the four tables in their connection with each other. The Analogies of Experience—notwithstanding the significance which Cohen himself attributes to the Anticipations of Perception—then show themselves to be the most important. Our analysis of a Newtonian nature showed that fundamental laws of Newtonian nature are here at issue. Kant carries out an axiomatic construction of theoretical physics. This axiomatic construction of physics is carried out under the direction of Kant's insight into the axiomatic structure of geometry and arithmetic. In all three systems, of arithmetic, of geometry, and of theoretical physics, we are concerned with a multiplicity of propositions which can be axiomatised, that is to say the totality of the propositions can be deduced from a part of them. The consideration which is decisive for Kant is that these systems cannot be deduced from the principle of contradiction alone, but that further determinations of a material kind must be added to it. The axiomatisability of the system then means that these additional determinations can be expressed completely in a small number of the propositions of the system. These propositions can be put first as axioms, so that all the other propositions of the system can be deduced from them. The axiomatisability of Newtonian physics is something to which nineteenth-century physics gave little attention, and consequently students of theory of science and interpreters of Kant also almost always passed it by. This is really remarkable in view of the fact that Newton himself used an underlying axiomatic structure. It was through the development of axiomatics in geometry, arithmetic, and logic that attention was turned again to the axiomatics of physics.[2] Understandably, the Newtonian axioms were treated with new respect, but the Kantian system of axioms was also studied with great interest. It is to be hoped that important results will soon be forthcoming in this field. Two points of view will presumably come to the fore. First the variability of the system of axioms, which we know so well from geometry. Which of the propositions in the system are those that should count as axioms, is by no means fixed; there can be a free choice, and sets of axioms for the same system of propositions can contain very different propositions. It is not improbable that in the axiom-systems of theoretical physics the same situation will arise as in the axiom-systems of logic, namely that there can be systems of axioms which consist of only one proposition; in this case, Newtonian physics would be deducible from one single proposition. This result would no longer have the significance, on the basis of our present-day knowledge of axiomatics, that it would have had earlier, but all the same it would be interesting enough.

A result such as this would make the formal structure of Kant's system totter, and it does not stand very firm in any case; but on the other hand another of Kant's fundamental ideas would probably prove to be fruitful. Kant connects the three Analogies with the three fundamental characteristics of time. There is in all probability a connection here which will prove to be a very fundamental one. Newtonian physics itself, as we can already see, is obviously most intimately connected with the concept of space and time that Newton uses.

There are two things that we can say. The Pure Principles, which we have seen to be the axioms of physics, are synthetic propositions and a priori propositions. At the first approach their synthetic character rests, as in geometry, on the fact that the totality of these propositions cannot be proved from the principle of contradiction alone. They contain new presuppositions which are additions and extensions to the presuppositions of logic, arithmetic, and geometry. This is true particularly of the law of causality which is an addition in this sense to the presuppositions of logic and mathematics. This would not be altered if anyone succeeded in finding a purely mathematical formulation of the causal law. His seeing that they have the character of extensions is regarded by Kant as his real philosophical achievement. We have seen that Kant finds himself in fundamental agreement with Leibniz, and we believe that Leibniz in fact conceived the principle of sufficient reason in this sense. It is from this point of view that attempts to prove the law of causality will have to be judged. I myself believe that attempts to prove the causal law will be successful and unsuccessful in the same sense as the attempts to prove the axiom of parallels. If a proof of the causal law is understood to be a proof from the principle of contradiction alone, a proof can scarcely be possible. A proof would still probably be impossible if all the methods of proof available in modern logic were admitted, and we know today that, contrary to Kant's opinion, logic can by no means be deduced from the principle of contradiction alone but contains further presuppositions. Proofs of the causal law are possible, but they can only be proofs from strong enough presuppositions, and this means that the causal law is proved from an equivalent law. Proofs of this kind are not only fruitful, they are very necessary; but proofs of this kind do not touch the unprovability of the causal law in the sense in which 'unprovability' was taken by Leibniz and Kant, namely in respect of the principle of contradiction.

We shall have to discuss further the a priori character of the Pure Principles and especially of the causal law; here we affirm only that the Principles are seen by Kant to be principles of possible experience and that only as such can they be proved. This approach can be well understood from the point of view of modern axiomatics. A set of axioms only has any meaning in relation to a system of propositions, and a 'proof' of a set of axioms is only possible in the sense of showing that it is an axiom-set for the system of propositions in question. In the same way the Principles are conceived by Kant as principles of possible experience, or quite concretely, the Principles are conceived as the axioms of classical physics.

The second of Kant's four tables is the Table of Categories. Here again Kant takes up the historical connections. He says explicitly that Aristotle was the first to look for these fundamental concepts: 'It was an enterprise worthy of an acute thinker like Aristotle to search for these fundamental concepts' (A 81, B 107). The question of the number and arrangement of the categories was then one of the great philosophical themes. Kant's table contains twelve categories (A 80, B 106):

TABLE OF CATEGORIES

1

Of *Quantity*
Unity
Plurality
Totality

2

Of *Quality*
Reality
Negation
Limitation

3

Of *Relation*
Of Inherence and Subsistence
(*substantia et accidens*)
Of Causality and Dependence
(cause and effect)
Of Community (reciprocity
between agent and patient)

4

Of *Modality*
Possibility—Impossibility
Existence—Non-existence
Necessity—Contingency

The old way of justifying the categories, which Kant in a sense retains, is to start from the assumption that the concepts which are in general use are composite. This composition can be resolved, with the result that the composite concept is analysed into its part-concepts, and any definition in terms of genus and specific differentia is an analysis of this kind. The composite concept 'man' is analysed in definition into the two part-concepts animal and rational. The concept 'force of gravity' is also composite and contains at least the concept of force as part of itself. The question then arises, whether this analysis of composite concepts can be continued to infinity or whether it has a finite conclusion. Leibniz answers, with Aristotle, that only the latter can be the case. Analysis of composite concepts would have to reach unanalysable fundamental concepts after a finite number of steps. These are then the ultimate elements and must be identical with the categories. All composite concepts can be made up by the combination of these elementary concepts and the construction of composite concepts becomes a purely combinatorial problem. This very brief sketch represents Leibniz's position, but its influence is also perceptible in Kant. He first calls the categories in general primary concepts (*Stammbegriffe*), but also uses the term elementary concepts (*Elementarbegriffe*) (B 109), which from Leibniz's point of view is the term to use. The idea of the construction of the system of composite concepts through the combination of fundamental concepts is also retained by Kant.

The categories when combined with the modes of pure sensibility, or with one another, yield a large number of derivative a priori concepts. To note, and where possible to give a complete inventory of these concepts, would be a useful and not unpleasant task, but it is a task from which we can here be absolved (A 82, B 108).

There may be room for doubt as to whether the system of derivative concepts is so easy to produce that to construct it can be called an easy and not unpleasant occupation, and in any case it must certainly be doubted whether this construction can be carried out by combining the categories with each other or with the pure modes of sensibility.

Now it is true that the structure sketched here would have to be called an essentially Leibnizian structure and that it could be called static. It is Kant's aim to free himself from a static conception and to arrive at a constructive conception of thought. He tries to interpret the categories as the fundamental acts of pure thought. But the difficulties in the relation between fundamental concepts and derivative concepts are only increased by a constructive conception. Kant speaks of subsuming the predicables of force, action, and passivity under the category of causality (A 82, B 108). But even if the concept of causality can be taken in a certain sense as an act of pure thought, how are the derivative concepts of force etc. to be taken as acts of thought? And even if the relation between categories and predicables can be understood as a relation between fundamental concepts and composite concepts, how is the corresponding relation between fundamental acts and derivative acts to be understood? That these problems are insoluble was shown well enough by the *Opus Postumum*. The fundamental difficulty is this, that we have here combinations which are to be analysed in a finite number of steps into fundamental elements which are not composite. The argument that a composite must have simple elements underlying it is certainly seductive, but it was Kant himself who brought to light the dialectical illusion in this idea, in the problem of the atomic structure of matter. It is certainly remarkable that in a way which is not at all easy to understand Kant falls victim to the same illusion in the problem of categories. Leibniz on the other hand is cautious to the last. It is true that Leibniz repeatedly demands a system of fundamental concepts which are not composites, but he is content to stay at the demand, and refrains from actually giving the system or at least from publishing it. Kant on the contrary not only gives the system of fundamental concepts in the table of the twelve categories, but he even calls this table his great achievement, with particular pride. It is pitiful to see how in the *Critique of Practical Reason*, the *Critique of Judgment*, and the *Metaphysical First Grounds of the Natural Sciences*, Kant forces all problems into the Procrustes' bed of the twelve categories. It is very improbable that this stark twelve should flow from genuinely Kantian presuppositions, and few scholars have assumed a real connection here. Indeed, it can be regarded as the common result of interpretations of Kant up to the present time, in particular those of German Idealism and the Neo-Kantians, that this rigid structure has got to be made dynamic and living.

The third table is the table of schemata, given in the chapter, Of the Schematism of the Pure Concepts of the Understanding. This chapter has always been notorious for its difficulty and obscurity and Kant himself says, 'This schematism of our understanding, in respect of its appearances and their mere form, is an art concealed in the depths of the human soul, whose real way of working we are hardly likely ever to discover from nature and to lay open to our gaze' (A 141, B 180). The inherent difficulties are admittedly increased by an ineradicable scientific prejudice. It is held to be a sign of a particularly scientific approach to treat the most general question immediately in every instance and then perhaps shamefacedly to quote concrete cases afterwards more or less as examples. But it must not be forgotten that this prejudice rests on certain opinions held in Greek logic on the relation between universal and particular concepts and that these presuppositions are not indubitable. Mathematicians at least have seen that it is not correct and not even possible to begin with the universal concept of number. A special class of numbers, namely the natural numbers, has to be treated first, and only from natural numbers can one ascend to the more general concepts of number, such as rational numbers, real numbers, and complex numbers. In the chapter on the Schematism Kant has made the argument more difficult by first giving the universal concept of the schema and then developing the schema of substance along with the other concrete schemata. In reality, however, if we are right, the problem of the schematism starts from the schema of substance. We therefore depart from the Kantian order and first treat the schema of substance, ascending afterwards to the general problem.

The category of substance is taken from spatio-temporal things. Substances, according to Aristotle, are men, animals, plants, elements (*Met.* IV.8.1017b10). Leaving out of account for the time being the question of the existence of God, all these substances exist at some point of time and in some position in space. By considering these substances, accidents can be distinguished, and from this point of view substance is what remains and is permanent in contrast with the changing accidents. Achilles is a substance and the wrath an accident, and in contrast to the wrath which swells and melts away, Achilles is what is continuously there. When formalised this leads to the principle of permanence: in all change of appearances substance is permanent. In this first respect therefore substance and accident have a temporal mode. But this temporal mode can be removed to give the pure relation between substance and accident, which only says that substance is the bearer of accidents and that the accidents are attributable to substance as determinations of it. This pure relation is from the side of substance a *subsistere*, from the side of the accidents an *inesse*, and Aquinas gives a definition which agrees with Aristotle, *esse accidentis est inesse* (*com. in met.* 894). This pure relation between substance and accident can be further considered in two ways. It can be considered purely ontologically: substance is then the substratum and accident the modification. But the relation can also be considered in respect of its logical form. The relation of substance and accident then corresponds to the relation of subject and predicate. Subject and predicate of the sentence 'Achilles is wrathful' show substance and accident separately. Ontologically regarded, substance is substratum, logically regarded, it is subject. And as accidents can be founded in other accidents or predicated of other accidents, substance must be defined as the ultimate substratum or the ultimate subject. Substance therefore shows two determinations, first as the permanent and then as the ultimate substratum or the ultimate subject, and the relation between these two determinations constitutes the problem of the schematism.

Substance considered as the permanent is the schema of substance.

> The schema of substance is the permanence of the real in time, that is, the representation of the real as the substratum of empirical determination of time in general, and so as abiding while all else changes. (The existence of what is changeable passes away in time, but time itself does not. To time, itself unchangeable and abiding, there corresponds in appearance what is unchangeable in its existence, that is, substance. Only in relation to substance can the succession and co-existence of appearances be determined in time) (A 143, B 183).

Besides this temporal determination of substance, which occurs in Kant as the schema of substance, there exists its pure function as subject.

> Now there certainly remains in the pure concepts of the understanding, even after elimination of every sensible condition, a meaning; but it is purely logical, signifying only the bare unity of the representations . . . Substance, for instance, when the sensible determination of permanence is omitted, would mean simply something which can be thought only as subject, never as predicate of something else (A 147, B 186).

Here substance is the subject which is not the predicate of something else. As this pure subject, substance is the pure category.

This is a particularly vivid example of what Kant means when he says at the end of the preface to the *Prolegomena* (IV,263) that the procedure underlying the *Critique of Pure Reason* is not analytic but synthetic. In an analytic procedure one starts from the fact that substances first demand consideration as temporally and spatially determined things. Accordingly permanence is then the first determination of substances, in contrast with the changing accidents. But permanence can be removed analytically, and substance can be further defined as ultimate substratum or as ultimate subject. In the synthetic procedure of the *Critique of Pure Reason* Kant goes the opposite way.

The pure category is first developed, then in the chapter on the Schematism permanence is added.

This temporal determination is then seen to be a necessary determination of substance. In order to apply the concept of substance at all, we have to go back to permanence. Unless it is thus determined in time, substance as a pure concept of the understanding has no object and no meaning. Conversely, however, this temporal mode of substance implies the limitation of the concept of substance to what is temporally determined. As the concept of substance only has any meaning for us in its temporal modification, we can only apply the concept to what can be temporarily determined.

> But it is also evident that although the schemata of sensibility first realise the categories, they at the same time restrict them, that is, limit them to conditions which lie outside the understanding and are due to sensibility . . . This meaning they [*sc.* the categories] acquire from sensibility, which realises the understanding in the very process of restricting it (A 146–7, B 185–7).

This connection between category and schema is most clearly apparent in the case of substance but it can also be seen in the other categories and their corresponding schemata and holds generally of the relation between category and schema. The categories are first found, in the analytic procedure, in a temporal modification, but abstraction can be made from this temporal modification to give the pure concept of the understanding. Conversely, in the synthetic procedure Kant develops the pure concept of the understanding first and then adds the temporal modification in the chapter on the Schematism, and only on the basis of this temporal modification can the pure principles be given. I believe that this taking apart in analysis and putting together in synthesis is not a mere juggling with parts but is a matter of showing up real connections. For Kant himself there is not first a pure category and then something added to it in the schema, but the temporal determination of the categories is something originally and inseparably given for us, which is merely analysed. Heidegger emphasised the importance that the problem of time acquires in this central position,[3] but we must refrain from pursuing this matter further. The decisive result for us is that the categories only have any meaning if they have a temporal modification and that this temporal modification of the categories in the schema at the same time limits the application of the categories to what can be temporally determined.

The first of the series of four tables is the Table of Judgments, with which the actual argument of the Transcendental Analytic begins. Judgments as a function of thought 'can be suitably represented in the following table' (A 70, B 95):

1
Quantity of Judgments
Universal
Particular
Singular

2
Quality
Affirmative
Negative
Infinite

3
Relation
Categorical
Hypothetical
Disjunctive

4
Modality
Problematic
Assertoric
Apodeictic

As the number of the categories, twelve, derives from these twelve kinds of judgments, the definitiveness and completeness of the table of categories depends on the definitiveness and completeness of the table of judgments. That is why, in the chapter heading, the table of judgments is called the transcendental clue to the discovery of all pure concepts of the understanding.

It can hardly be maintained that Kant justified the large claims that are made by this choice of method, and even Klaus Reich's very comprehensive treatment of the subject [4] does not seem to have produced a generally acceptable result. Criticism is directed to two points, whether Kant himself proved that his table was complete and whether on the method which Kant has chosen a proof of its completeness is possible at all. Hegel made this point clearly. He says of the table of categories, and what he says must also hold of the table of judgments on which it is founded:

> In the usual logical treatises various divisions and kinds of concepts occur. It is immediately obvious that it is inconsequent to introduce the kinds like this: there are the following kinds of quantity, the following kinds of quality

etc. This gives no other justification than that these kinds occur and show themselves in experience. We arrive in this way at an empirical logic—a strange science, irrational knowledge of the rational (V,52).

There can be no serious doubt that Hegel's assertion is correct, and Kant himself introduces the table of judgments with the laconic words 'we find that . . .'. Kant himself at all events gave no proof of the completeness of the table, and if any such proof is offered, it must always remain doubtful whether it really is a Kantian proof.

It must indeed be asked whether on Kant's chosen method a proof of the completeness of the table is possible at all. The intention is to construct the first table, the table of judgments, by purely logical means, and the underlying idea is no doubt that formal logic, in contrast with mathematics and metaphysics of nature, is an analytic science. But the table of categories is essentially a problem in an indisputably synthetic science, and some essential factor or factors must therefore be added to give to this science the character of synthesis. If formal logic is to be a purely analytic science, it is not easy to see how the essential content of the Transcendental Analytic could flow from a purely formal logic. Two possibilities could be conceived. Either formal logic is not a purely analytic science, or the table of judgments does not flow from purely formal logic but from some other logic, perhaps a transcendental logic. Klaus Reich has contributed a number of important considerations to this difficult topic but does not seem to me to have reached a definitive solution.[4]

The latter solution, to take the table of judgments not from formal logic but from a transcendental logic, seems to be considered by Kant himself, for instance when he says, 'In like manner infinite judgments must in transcendental logic be distinguished from those that are affirmative, although in general logic they are rightly classed with them and do not constitute a separate member of the division' (A 71, B 97). But the assumption that there is a transcendental logic has to meet great difficulties. The title 'transcendental logic' is already booked. The Transcendental Logic is complementary to the Transcendental Aesthetic and contains the Analytic and the Dialectic. But apart from this difficulty, there is no firm handhold in Kant's own thought for the assumption that there is a special transcendental logic. The other assumption, that formal logic is not a purely analytic science, is equally difficult. To go back to Kant's connections with Leibniz, both logic and mathematics are analytic for Leibniz in the sense in which Kant calls judgments analytic, namely because all true propositions in both disciplines can be proved from the principle of contradiction alone. As we have seen, Kant holds that this theory of the foundations of mathematics is incorrect and that the propositions of mathematics are synthetic judgments. But from this point of view one has to say that Kant retains the analytic character of formal logic in the sense specified, and everything that Kant says on this point agrees with this. This is the objection that Cohen raises and Klaus Reich does not seem to have been able to avoid it. Cohen says:

> The forms of thought cannot be taken directly from the kinds of judgment which formal or general logic distinguishes; for these kinds of judgment are mere thought-constructions and the judgments are analytic. What we are looking for is the forms as such of the thinking of synthetic judgments. The unity of consciousness, for which thinking has to provide the means, is a synthetic unity of consciousness. Thus the forms of this synthetic thinking cannot be taken from the kinds of analytic thought.[5]

Cohen further recognised quite clearly and said equally clearly that the system of four tables was not originally taken from formal logic at all. 'Kant developed Newton's principles into his own synthetic principles'.[6] Even though we are still without an exact and detailed foundation for this view, such as could be achieved by a close investigation of Kant's thought between 1750 and 1780, I hope that in what I have said I have given further plausibility from the systematic standpoint to Cohen's attitude.

For our interpretation this problem of the derivation of the table of categories from the table of judgments is of secondary importance. For the ontological view at which we are aiming, the number and nature of the particular categories is not important. The ontological attitude of transcendental idealism does not depend on whether there are twelve categories or twenty-four.

Furthermore, the development of modern logic has shown that Kant's fundamental distinction between mathematics and logic cannot stand. This development has by no means brought about a refutation of Kant; we have indeed drawn on the results of modern logic extensively for our interpretation. The development has in our opinion shown that Kant did not go far enough. Kant recognised the synthetic character of mathematics but retained the analytic character of logic; I should like to say now in a Kantian sense that logic is also a synthetic science. Cohen's view that the four tables are to be regarded as an inner unity would thereby be confirmed.

We have briefly considered the system of the four tables, the tables of judgments, of categories, of schemata, and of principles. How far are the fundamental laws of nature proved by this system of four tables, and in particular how far is the causal law proved by the system?

The proof as Kant gives it depends essentially on the completeness of the tables. Nature, as analysed by Kant, does not mean things, natural objects themselves, but conformity to law as such; it therefore means here nature as it appears and is known primarily in theoretical physics. On the other hand the four tables comprehend the complete scope of all our ways of knowing. It follows directly that all our knowledge of nature must be formulated with the ways of knowing which are completely laid down in the four tables. As the totality of these ways of knowing includes only one possible way of knowing temporal sequences, namely the causal law, it follows in particular that all temporal sequences according to law have to be presented by means of the causal law. This means that the causal law has a priori validity for the presentation of all temporal sequences. The same holds universally for the principles as a whole. As we have no other ways of knowing at our disposal, the fundamental conformity of nature to law must be present in these principles, and the principles are therefore valid a priori.

The problem is perhaps made even clearer by weakening Kant's actual argument in a certain way, and considering the problem purely as one of presentation. The system of the four tables then contains the totality of possible kinds of presentation and it is thus certain that every presentation of nature must be included in the totality of possible kinds of presentation.

The inference is certainly binding. If the totality of all ways of knowing or means of presentation can be specified, then a pronouncement can be made about every actual piece of knowledge and every actual presentation. But it is also plain to see what a burden of proof Kant has here taken upon himself. We have seen that both Aristotle and Leibniz had such a system of fundamental concepts in view, but both of them carefully avoided asserting that they could actually give a complete system of fundamental concepts.

Kant on the other hand says repeatedly that reason can be known completely. In this statement, even if it be restricted to reason's knowledge of itself, there is the characteristic but scarcely justifiable optimism of the Enlightenment. Why should reason be able to know itself completely? Developments since Kant's day have not confirmed Kant's idea of the a priori in this stark formulation. The Kantian formulation contains the thesis that the causal law is the only possible law of temporal succession, and, with respect to the totality of principles, it contains the thesis that Newtonian physics is in respect of its fundamental laws the only possible physics. Even though the question of the validity of the causal law cannot yet be regarded as definitively solved, it is the idea of making an a priori pronouncement about every actual law on the basis of the totality of all possible laws, and about every possible presentation on the basis of the totality of all possible kinds of presentation, that constitutes the decisive Kantian approach. Systematic philosophy will have to pursue this approach further. From the fact that Kant attributed too much to the concept of the a priori it does not follow that the concept of the a priori has no meaning at all. The Marburg school saw that the real problem for systematic philosophy is to reformulate the relation between the a priori and the a posteriori.[7]

An important contribution to this question was made by Hugo Dingler and C. F. von Weizsäcker,[8] when they recognised the a priori status of Newtonian physics in relation to quantum physics. Dingler was the first to see that every statement in quantum physics must be reached through a statement in classical physics. If this could be demonstrated universally, then classical physics would indeed stand as an a priori to quantum physics. A way of considering the nature of an a priori could perhaps be found by considering Aristotelian logic as the foundation of all other logics, Euclidean space as the foundation of all other spaces, and classical physics as the foundation of all other kinds of physics.

Kant on the contrary maintains, in all probability going far beyond what can be proved, that Newtonian physics in its fundamental conformity to law is the only possible physics, and there follows then directly from this the a priori character of these fundamental laws as they are formulated in the Categories and the Principles.

§ 13

The Relational Character of Newtonian Nature

We saw in the first chapter that it is a matter of fundamental importance that space is a complex of relations and that mathematics only treats of relations. The same importance attaches to the relational character of nature. The relational character of nature, and thus of theoretical physics, is an immediate consequence of the relational character of mathematics. The statements of theoretical physics can be presented in mathematical form and the relational character of mathematics is thereby carried over into physics.

Kant's insight into the relational character of nature is very important for him, although he did not entirely succeed in welding this problem into the systematic argument of the Transcendental Analytic. It found a place in the appendix, under the title of The Amphiboly of the Concepts of Reflection. The problem is presented partly in this appendix itself and partly in a discussion of Leibniz which is further appended.

The relational character of nature permeates all the determinations that belong to pure nature, including the concept of substance itself.

> The Inner and the Outer.—In an object of the pure understanding, that only is inward which has no relation whatsoever (so far as its existence is concerned) to anything different from itself. It is quite otherwise with a *substantia phaenomenon* in space; its inner determinations are nothing but relations and it itself is nothing whatever but a totality of relations (A 265, B 321).

What Kant says here about substance in nature is also valid for nature as a whole. Nature itself is nothing whatever but a totality of relations. This is in fact the underlying attitude of the new natural science but it means a complete break with the Greek attitude. The new natural science treats of relations, Greek natural science treats of the essence and the essential qualities of living things. In this sense Greek natural science enquires after the essence and the essential qualities of animals and plants and of the elements. Only from this standpoint can a deduction of the elements, for instance, be attempted, as it was by both Plato and Aristotle with the same aim though in different ways. The Scholastics recognised early that this aim, knowledge of essence and of essential qualities, had not in fact been reached in Greek natural science. But it was also seen very early, for example by Duns Scotus, that to set this aim for the natural sciences raises difficulties of principle. Does the physicist really recognise the essence of water, does he even recognise the essence of water in any approximation? Even the findings of Greek physics do not support this view. The new natural science renounces at all events knowledge of essence and knowledge of essential qualities; it is consciously a science only of relations. Kant expressed the standpoint of the new natural science particularly clearly. He goes back again to the problem of spatial substance. 'For, as we find, an abiding appearance in space (impenetrable extension) can contain only relations and nothing at all that is absolutely inward, and yet be the primary substratum of all outer perception' (A 284, B 340). This renunciation of knowledge of essence and limitation to relations constitutes the peculiar pathos of modern natural science and was also expressed convincingly by Kant.

If the complaints that we have no insight whatsoever into the inner nature of things mean that we cannot conceive by pure understanding what the things which appear to us may be in themselves, they are entirely illegitimate and unreasonable. For what is demanded is that we should be able to know things, and therefore to intuit them, without sense, and therefore that we should have a faculty of knowledge entirely different from the human, not only different in degree but also different as regards intuition and in kind—in other words that we should be not men but beings of whom we are unable to say whether they are even possible, much less what they are like. Nature's inner recesses are penetrated through observation and analysis of appearances, and no one can say how far this knowledge may in time extend (A 277, B 333).

The modern natural scientist does not feel the limitation to appearances as a restriction, and Kant could therefore answer Haller's well-known complaint,[1]

> Nature's interior parts are not for creature's sight,
> Too happy to whom she shows merely her outer shell,

thoroughly in the way that any modern scientist would answer it. On the other hand Goethe's continuation of the same theme,[2]

> Nature has neither shell nor kernel,
> Nature is everything at once,

would take us into further realms.

The thesis that nature is relational in character, like the thesis that space is relational in character, is in a certain sense a result arrived at purely by categorial analysis, but it is none the less most intimately connected with the ontological characterisation of nature. The relational character of nature is probably one of Kant's deepest grounds for declaring nature to have the character of appearance. In this way the same system of connections could be found in the case of nature as in the case of space. The objects of natural science not only consist of nothing but relations, remarkable as this may sound, but they *must* consist merely of relations because they are mere appearances. 'It is certainly startling to hear that a thing is to be taken as consisting wholly of relations, but such a thing is mere appearance . . .' (A 285, B 341). Conversely the relational character of nature entails the thesis that nature is phenomenal.

§ 14

The Transcendental Ideality of Newtonian Nature

The theses that nature is a priori and relational in character, which are in a certain sense theses in categorial analysis, are followed by the ontological thesis that nature has the character of appearance. That nature has the character of appearance is what the Transcendental Deduction is about, and as it is formulated by Kant in so many ways and in so many places it will be enough to mention a few of them, for instance, 'But when we consider that this nature is nothing but a totality of appearances and not a thing in itself. . . .' (A 114). The most striking and hence historically most effective formulations are those in which Kant speaks of the understanding prescribing laws to nature. 'Thus the order and regularity in the appearances which we call nature, we ourselves introduce, and we could never find them there had not we ourselves, or the nature of our mind, originally set them there' (A 125). Again, put rather differently,

> However exaggerated and nonsensical it may sound to say that the understanding is itself the source of the laws of nature, and so of the formal unity of nature, such an assertion is none the less correct and is in keeping with the object, namely experience (A 127).

These examples are from the first edition. The Transcendental Deduction is altered in the second edition, and these alterations constitute the essential difference between the first and second editions of the *Critique of Pure Reason*. Our fundamental ontological thesis is not altered but only brought out even more clearly. 'Categories are concepts,' says Kant in the second edition, 'which prescribe laws a priori to appearances and therefore to nature, the totality of all appearance (*natura materialiter spectata*)' (B 163).

The importance of this thesis emerges particularly clearly if the *Critique* is compared with the *Inaugural Dissertation* of 1770 *De mundi sensibilis atque intelligibilis forma et principiis*. We have already seen that in respect of space and time Kant takes up in 1770 the standpoint of the *Critique of Pure Reason*, in that he teaches the phenomenality and thus the transcendental ideality of space and time. We also saw how this is a development of Leibniz's position. In respect of the problems of the concept 'world' also, one can say that in the *Dissertation* Kant had to a large extent reached the position of the *Critique*. In 1770 Kant had seen the problem of antinomies and its ontological consequences. But in respect of the categories the position which Kant takes up in 1770 is exactly opposed to that of the *Critique*. 'Things which are thought sensitively [*sensitive cogitata*] are representations of things as they appear, intellectuals [*intellectualia*] on the contrary of things as they are' (§ 4,II,392). The pure concepts of the understanding include in particular the causal concept and the concept of substance, and the concepts of cause and of substance therefore represent things as they are in themselves. The *Critique* breaks in principle with this view. The concepts of cause and of substance present things not as they are in themselves but only as they appear to us.

In our interpretation of the fundamental ontological thesis that nature is only appearance, we shall first try to understand it from a present-day viewpoint, then briefly investigate how Kant provides a foundation for it, and finally bring out its significance by comparing it with other possible theories.

The Kantian thesis that nature has the character of appearance can be understood, we believe, on the basis of the modern view in theory of science that nature has the character of a model, or more exactly, that modern scientific ideas have the character of a model. The thesis that the concepts used in the natural sciences, and in physics in particular, have the character of models has rightly acquired great significance. When the phenomena of electricity and magnetism were investigated experimentally, the question of a theoretical grasp of them arose. Understandably the first question was whether electrical phenomena can be presented as a purely mechanical process. Even after Maxwell had found an independent theoretical presentation of the phenomena, the question was not dropped but took the form of whether it was possible to reduce Maxwell's equations to mechanics. A presentation of electrical phenomena by purely mechanical means would be a mechanical model, and it is very significant that such a model could not be found. The concept of a model acquired an even greater significance in quantum physics, when Niels Bohr created what has come to be called Bohr's model of the atom. In Bohr's model the atom is considered as a mechanical planetary system and is thus understood on the basis of solar systems. This model of the atom also brings out the great significance of the concept of a model as such. We know for certain that the atom cannot be a purely mechanical system, and we know for certain that this picture of a solar system is an inadequate presentation of the atom. None the less, this inadequate and inaccurate picture presents the atom in a fruitful way and leads to a wealth of genuine knowledge about the atom. It was these features especially of the model of the atom that won universal recognition for the idea of a model. Scientists have convinced themselves that scientific ideas neither fit things adequately nor can they fit them adequately. All our scientific representations are only models, they are only presentations of natural events, and it is not necessary that the events in nature should be exactly as they are presented in the models.[1]

We believe that by starting from the concept of a model we can approach Kant's thesis that nature has the character of appearance. We have shown that by nature Kant means a Newtonian nature. What is nature, if it is only a Newtonian nature, that is, if it only contains masses which move in a way that is strictly determined according to Newtonian laws? Kant's answer, if we have understood it correctly, is that Newtonian nature is only a model. With the concepts of pure mechanics we can give a presentation of nature that fits the events, but it does not follow from this that the events in themselves are purely mechanical events; and with Bohr's model of the atom we can give a presentation of the atom which fits it in many respects, but it does not follow from this that the atom itself is a system after the manner of a solar system. What Kant means is that nature appears to us as a purely mechanical system, we represent it to ourselves as a purely mechanical system, but it does not follow from this, or at least it need not follow, that nature, taken as nature in itself, is really a mechanical system.

Kant's proof that nature has the character of appearance is dependent on space and time and the world having the character of appearance. This means that the foundations of transcendental idealism are provided in the Transcendental Aesthetic and the Transcendental Dialectic, while the ontological determination of nature draws on the results reached in the Aesthetic and the Dialectic. This is the reason why, following Kant's personal development, we put our discussion of the Dialectic before that of the Transcendental Analytic.

The Kantian proof consists in showing that all the concepts of pure natural science contain spatial and temporal components, thus that Newtonian nature is a spatio-temporal system, and thus that because space and time have the character of appearance, nature as a whole has the character of appearance. The demonstration that temporal factors are contained in all pure principles is actually given in two places, in the chapter on the Schematism and in the Transcendental Deduction.

We have already seen that in the chapter on the Schematism it is shown that the pure concepts can only be applied if they are schematised in time, that the concept of substance, for instance, can only be applied when it has been schematised as the concept of permanence. Through this temporal determination the concept of substance is made capable of being applied, but it is thereby also restricted to temporal events.

> The categories, therefore, without schemata, are merely functions of the understanding for concepts, and they represent no object. This meaning they acquire from sensibility which realises the understanding in the very process of restricting it (A 147, B 187).

Causality and substantiality are thus restricted to temporal events and are thereby submitted to the character of appearance which belongs to all temporal events, and what is true of these, the two most important fundamental concepts, is also true of the totality of fundamental concepts. Newtonian nature treats of temporal and spatial events and therefore treats of appearances.

Similarly the Transcendental Deduction also establishes an original connection between the pure concepts of the understanding and the pure intuitions of space and time. Kant summarises this as clearly as one could wish.

> Brief outline of this deduction—The deduction exhibits the pure concepts of the understanding (and hence all theoretical a priori knowledge) as principles of the possibility of experience; it exhibits experience as the determination of the appearances in space and time in general; and finally it exhibits this determination as following from the principle of the original synthetic unity of apperception, which is the form of the understanding in its relation to space and time which are the original forms of sensibility (B 168).

It cost Kant a lot of trouble to reach the position that nature also has the character of appearance. The way from the *Dissertation* to the Transcendental Analytic was hard and difficult. This is particularly easy to understand if we consider that it would have been easy to draw quite a different conclusion from the completeness and uniqueness of the table of categories. We have seen that the completeness of the table of categories involves the assertion that Newtonian nature is the only possible nature. The assertion so willingly made in the nineteenth century, that all physics is mechanics, in fact also underlies Kant's position. In terms of our interpretation we can formulate it in this way: for a nature according to law there is only one possible model, namely Newtonian nature. It would now be easy to go on to the inference: if Newtonian nature is the only thinkable nature, nature in itself must also be a Newtonian nature. With space and with the concepts of a world it was different. Other spaces, perhaps very many other spaces, were thinkable, and as regards the concepts of a world at least two different possibilities were thinkable, finiteness and infinity, atomic theory and theory of a continuum. Under Kant's presupposition (which, as we now know, is incorrect) only one kind of conformity of nature to law is thinkable. In spite of this Kant comes to the correct conclusion, taking his lead from the original connection between pure concepts and pure intuitions. By this means in the arduous years from 1770 to 1780, Kant discovered that a purely mechanical nature has the character of appearance.

This theory is decisively opposed to two other attitudes, the theologically grounded attitude and the purely materialistic attitude. On the theologically grounded attitude, as we have chosen to call it, natural science, when it reaches true knowledge, is considered as insight into God's plan of creation. It is represented by Plato, Augustin, Leibniz. We have seen how strong are the objections to such a theory especially on the basis of the antinomies and also of the problem of space and time. On the materialistic attitude a grounding of truth in God's thinking is dispensed with, but it is at least assumed that nature really is as it is conceived in mechanics, that is, that nature itself is a purely mechanical system. In opposition to this, Kant asserts that the conception of nature that is gained in pure mechanics is essentially determined by human conditions, that it is improbable that God sees the world in the form of pure mechanics, and that it is not even certain that all thinking beings have to think in this way. It cannot be demonstrated that nature in itself is a Newtonian nature, and we have to reckon with Newtonian nature being only the way in which we human beings represent nature to ourselves.

CHAPTER IV

UNITY

§ 15

Unity in the Ontologies of Plato and Aristotle

WE have attempted to arrange the problems of the Transcendental Analytic into two groups, the problems of nature, which are in some ways regional, and the universal problems of unity in general. In bringing forward the problems of unity for discussion in the Transcendental Analytic, Kant goes outside his division of the *Critique of Pure Reason.* The theme of the Aesthetic, namely space and time, and the themes of the Dialectic, in particular the world, are all unities. The unity which we are now investigating is not the category of unity which appears in the table as the first of the categories. This can easily be seen by considering that the table is itself a unity. There must be a highest unity, from which the pure intuitions of sensibility, the pure fundamental concepts of the understanding, and the transcendental ideals of reason can all be understood as unities. Kant himself says:

> The representation of this unity cannot, therefore, arise out of the combination. On the contrary, it is what, by adding itself to the representation of the manifold, first makes possible the concept of the combination. This unity, which precedes a priori all concepts of combination, is not the category of unity (§10) [i.e. §10 of the *Critique*]; for all categories are grounded in the logical functions in judgments, and in these functions combination, and therefore unity of given concept, is already thought. Thus the category already presupposes combination. We must therefore look yet higher for this unity (as qualitative, §12), namely in that which itself contains the ground of the unity of diverse concepts in judgments, and therefore of the possibility of the understanding, even as regards its logical employment (B 131).

In this ascent to a supreme concept of unity Kant reaches the climax of his thought and at the same time establishes his real connection with the great traditions of philosophy. At the head are the Greek physiologers with their great generalisations, led by Thales with his 'everything is water'. Here we have a most intimate intermingling of science, cosmology, and ontology. The proposition has first a scientific (in a sense, chemical) meaning, everything consists of water. Its cosmological aspect is perhaps easiest to understand from the romantic attitude: everything is fluid, everything is mobile, real existence is to be found among the things that have the lightest and most lively mobility. But it also contains an ontological assertion, namely everything is one. The world is one, and this is because it is made of one material, namely water. It was Aristotle who made it clear that the ontological thesis that all is unity is concealed in Thales' 'everything is water', and it is the ontological content of these propositions that he is summarising when he speaks of 'Those, then, who say the universe is one, and posit one kind of thing as matter, and as corporeal matter which has spatial magnitude . . .' (*Met.* I.8.988b22). Thales' proposition contains the thesis that the world is one, but it also contains a pronouncement about the manner of this unity—the unity of the world consists in its being made of one material.

It is possible to consider the early history of Greek thought as an attempt to clarify this proposition about the unity of the world and the essence of this unity. Parmenides might stand half way between an intuitive attitude and a purely conceptual formulation. It must not be overlooked that the being of Parmenides still contains important intuitive elements. It is 'completed in all directions, like the mass of a well-rounded sphere, equally great from the middle in all directions' (fr. 8, 42–4). These intuitive elements are not intended to be taken as a simile but appear again and again and must be given their full meaning. But Parmenides then reaches the first conceptual determinations of being, among them the fundamental determination, more important for us than any other, that being is one (B 8, 6). The great fundamental thesis of transcendental philosophy appears here for the first time in its full clarity; being is unity, *ens et unum convertuntur*.

The philosophy of the Pythagoreans forms the next important stage of the development. Aristotle devoted a special work to the Pythagoreans. If we still had it or could reconstruct it fairly completely, we should be able to gain an appreciably deeper insight into the history of philosophy. We only have the kernel, in Aristotle's account in the first book of the *Metaphysics*. We have to understand the philosophy of the Pythagoreans in two stages.

In the first stage numbers are assumed to be the principles of all that exists, and in the second stage the question is raised, whether the numbers themselves have to be derived in their turn from principles. The even and the odd, the limited and the unlimited, are then said to be principles (*arche*) of numbers. The number one consists of both and hence is both even and odd. Thus Aristotle can say that for the Pythagoreans number proceeds from unity, and the whole heaven consists of numbers (*Met.* I.5.986a15–21).

Aristotle's account can be adequately supported by the few surviving fragments. Archytas calls number and quantity the two original forms of being. Philolaos praises the power of numbers: 'The effectiveness and the essence of numbers must be judged according to the power that lies in the number 10. For it is great, all-completing, all-effecting, and the *arche* and the leader of all divine, heavenly, and human life' (fr. 11). Number in general and the number 10 in particular are here called the *arche* of all things, but in another passage Philolaos also calls unity itself the *arche*. 'One is the *arche* of all things' (fr. 8). For the Pythagoreans numbers are thus the principles of the world, and the limited, the unlimited, and unity (combined in a way that is not entirely clear) are in their turn the principles of numbers.

The first stage of this Pythagorean doctrine, in which numbers are said to be principles of the whole, can still be imitated intuitively. The Pythagoreans discovered the great significance of number in music. Only those notes which are numerically determined are notes. A sound that lies outside these intervals is not a note. And not only musical notes but the whole course of the world, as it is given by the course of the sun, the moon, and the stars, is numerically determined in its articulation and its existence. The deep joy that this insight gave to the Pythagoreans drove them far beyond these bounds to the universal assumption that everything is number.

The proposition that everything is number, which transcends all bounds, retains its intuitive connection with the propositions of the physiologers, and it is not a very far step from the proposition 'all is water' to the proposition 'all is number'. As for Thales everything consists of water, so for the Pythagoreans everything now consists of numbers. This means further that the numbers exist of themselves, that, preceding things, they are contained in things as their real content, and the essential point for us is now this precedence of numbers. We have too few texts to support this exactly in detail, but what we have must, it seems, be understood to mean that numbers precede things in the sense that numbers would still exist even if there were not any things. It is not that things exist originally and the numbers are then the numbers of things, but numbers exist originally, and from numbers there arise things.

This relationship of numbers, as origin, to things, as what has originated from them, holds again in a deeper sense of unity, both in the relation of unity to numbers and in the relation of unity to things. Unity is the origin of all things, and this is true in the more emphatic sense that unity precedes things, that numbers and things originate from unity, and that there could be unity without things but never things without unity. Aristotle interprets this fundamental ontological approach of the Pythagoreans to mean that they regarded unity as a substance (*Met.* X.2.1053b12). According to Aristotle's conception, unity is a substance for the Pythagoreans because, being a principle, it can exist of itself, without things. Aristotle's account makes the Pythagorean doctrine of numbers and of unity both historically and systematically the introduction to Plato's doctrine of ideas, as was in fact the case.

What Plato's doctrine of unity was, is not so easy to say. The doctrine of ideas occurs in two different forms, as transcendent doctrine of ideas and as dialectic. By a transcendent doctrine of ideas we mean a doctrine of ideas in which an independent existence, intuitively an existence above the heavens, is attributed to ideas, as in the *Phaedo* or the *Timaeus*. There is no doubt that in the *Parmenides* and the *Sophist* such a transcendent doctrine of ideas is criticised extremely sharply, but this only makes the presence of the two forms of the doctrine in Plato's works even more puzzling. In the Academy itself, presumably under Plato's eyes, there was discussion about how the relation of 'the mathematical' (numbers and figures) to the ideas was to be conceived. These discussions do not seem to have been put down independently in writing, and all that we know significantly about them comes from Aristotle's reports. In the main, two views seem to have been formed. Confining ourselves to numbers, from the one point of view numbers were regarded simply as ideas, while from the other point of view numbers were given an intermediate existence between the ideas and things of the senses. The problem of unity, with which we are concerned, is made still more difficult by the fact that for the Greeks one is not really a number, two being taken to be the smallest number. This indeterminateness of the doctrine of numbers is only the expression of the mobility of Platonic thinking, and any interpretation has to decide at this point what to keep and what to leave out. From our point of view we may keep to Plato's presentation of the one and the good as the archetypes of the ideas, and in the *Republic* (524e) Plato speaks of the great question as to what the one is.

Unity is thus an idea, and the question now is, what is the ontological significance of calling unity an idea? We take our lead from Aristotle, and his account in *Metaphysics* I.6 may be freely reproduced in the following way. Plato posits the objects of mathematics, namely numbers and figures, as a third region of being besides the things of the senses and the ideas. Numbers and figures are differentiated from things of the senses in being timeless and immobile. But they are differentiated from the ideas in that of every number there are many alike, as there are three fives in the equation $5 \times 5 \times 5 = 125$, but of each idea there is only one, and only one idea of the good. But as the ideas are the principles of all other existents, Plato believed that the elements of the ideas must be the elements of all existence. Now Plato assumes that there are ultimate elements of the ideas, namely the great and the small as matter, and unity as substance, since he tries to understand ideas as consisting of the great and the small by means of their participation in the one. That unity itself has its own substantiated being, and that the one is not the name for an existent determined as to its content in some other way, is a view which he shares with the Pythagoreans, as is also the view that the numbers are the real cause of the existence of all other things (*Met.* I.6.987b22).

The real core of this Aristotelian interpretation is that Plato defined unity as substance, and Aristotle says this in the passage already quoted: '. . . whether we must take the one itself as being a substance (as both the Pythagoreans say in earlier and Plato in later times)' (*Met.* X.2.1053b11).

There has been much controversy about whether this criticism of Plato is not a complete misunderstanding, but from our point of view it seems to be possible to avoid the negative attitude and give a positive interpretation of Aristotle's criticism. Aristotle poses the question, what does it mean if unity or if the good is called an idea? The legitimacy, indeed the necessity of such a question cannot be disputed. When Aristotle says that Plato called unity a substance what he means is that although Plato said that unity was an idea, he only attributed to this idea the existence that substances have. This argument runs like a crimson thread through the whole criticism of Plato in the

Metaphysics and is known as the problem of duplication. It is usually considered as a question of content. The objection that there is a duplication means for instance that as well as particular men the idea of man is posited and that this idea does not give anything new but only the particular man posited a second time, nothing more than a duplicate of him. But the same argument can also be regarded as a question of wholly formal ontology. It then means that ideas have no other existence than phenomena, the existence of ideas is only the existence of phenomena posited a second time. In our concrete case, unity taken as an idea has no other existence than substance which impinges on the senses, unity itself is in the last resort said to be only a substance. This objection is always taken purely negatively, but that is not at all necessary, and this criticism of Plato can be made to yield a very positive lesson. There is no doubt that both for the Platonic doctrine of ideas and for the Aristotelian ontology, unity is not a substance and not a thing. The fundamental argument in Aristotle's criticism must then be taken to mean that one must in fact distinguish between the status of unity and the status of things, that Plato, in distinguishing unity as an idea from things as phenomena, intended to make such a distinction and in fact made one, and the criticism would then be that Plato did not succeed in distinguishing the status of ideas from that of phenomena in a clear and well-grounded way. Aristotle's objection would then be not that Plato distinguished two modes of being, but that he did not succeed in making such a distinction really clear and in the existence of ideas only posited the existence of phenomena a second time. But in reality the status of ideas is different from that of phenomena. Being, to use Aristotelian terminology, is not a univocal concept; the concept of being unfolds into a manifold and this manifold was seen for the first time in the distinction between ideas and phenomena. With every justification Aristotle concludes his criticism of Plato in the first book of the *Metaphysics* with an emphatic allusion to the unfolding of the significance of the concept of being. 'If we search for the elements of existing things without distinguishing the many senses in which things are said to exist, we cannot succeed' (*Met.* I.9.992b18).

The distinction which Aristotle demands leads, it is true, to the most fundamental and most difficult problem of metaphysics, the determination of the different modes of being, in particular cases to the distinction of the existence of space from the existence of spatial things, and the existence of unity from the existence of unitary things.

We now see that all concrete problems issue in the problem of unity. We had to distinguish the existence of space from the existence of spatial things, the existence of time from the existence of temporal things, the existence of the world from the existence of things in the world, and the existence of nature from the existence of natural things, and now it is the existence of unity which is to be distinguished from the existence of all the unitary things. Everywhere we find the same problem: ontological exposition follows first the tracks of linguistic exposition. The ontological exposition starts from the fact that in all these cases language operates with substantives, speaking of space, time, world, nature, unity. Through this substantification language puts together space, time, world, nature, unity, existence, sun, moon, tree, mountain, man, and this is what is intended in language. Ontology follows language in its first approach and out of the substantification makes a hypostatisation. In its first approach ontology assumes that space, time, world, nature, unity are like the things which are arranged in them. The first ontological interpretation thus makes these objects, in particular space and unity, into substances.

From the Aristotelian point of view the doctrine of ideas signifies the attempt to break away from this first approach. When the one and the good and the like are taken to be ideas, what this means is that unity and likeness are something different from things which are alike and things one by one. But Plato did not succeed in showing the difference in their ontological status; ideas were certainly supposed to be something different from things but, says Aristotle, if one takes what Plato actually said about them, the ideas are only things, only substances again.

From this point of view Aristotle's ontology would be an attempt to recognise the fundamental problem of the doctrine of ideas and to take it further. This continuation then produces the Aristotelian doctrine of the plurality of the modes of being. Being is not a unitary, univocal concept, but is only an analogical concept.

The variety of the modes of being shows itself in various different directions. First in the series of categories: being a substance, being a quality, being a quantity, and being a relation, have being each in a very different way. Then, in another dimension, *dynamis* and *energia*, potency and act; and in another dimension again, the transcendental determinations, being as such, being one, being perfect, and being true. The doubleness of being assumed by Plato develops into something much richer in the Aristotelian ontology. What is specially important, both systematically and historically, is its development into a multiplicity of transcendental determinations.

We give an account of the Aristotelian doctrine of transcendentals in the form in which it was given by Aquinas, considering the Thomistic ontology to be a genuine continuation of the Aristotelian ontology.

§ 16

Unity in Scholastic Ontology

The subject of unity is treated by Aquinas in Quaestio 11 of the first part of the *Summa Theologica*, and this quaestio deserves to be reckoned among the classical enquiries into the problem of unity. It treats the unity of God, and it is a matter of fundamental importance that the problem here is the unity of God and not the unity of for instance a genus. In Aquinas's enquiry the transfer of the problem to the theological field, on which we have already touched in § 10, has taken place, but in spite of the change in interest the unity of the problem and the unity of the answer are still apparent.

The treatment of the unity of God in this quaestio is divided into four articles, (1) Whether *one* adds anything to *being*? (2) Whether *one* and *many* are opposed to each other? (3) Whether God is one? (4) Whether he is in the highest degree one? (*s.th.* I,11). For our enquiry the first and the third articles are of particular importance. In the first and third objections to the first article Aquinas first expounds the problem in the following way. The first objection takes its start from the possibility of considering unity as a categorial determination, with the consequence that unity must add something to being: *ergo unum addit aliquid supra ens*. In his answer to this objection Aquinas points out that a distinction must be made between transcendental unity and categorial unity. Rickert found a happy phrase for it by saying that we must distinguish between unity and the one.[1] The ignoring of this distinction, Aquinas goes on, leads to incorrect interpretations, such as those of Pythagoras and Plato, who maintained that unity and the numbers are substances. This thesis rests on the Aristotelian interpretation of the doctrine of ideas which we have already discussed and is presented in full by Aquinas in his commentary on the appropriate passages of Aristotle's *Metaphysics*. Thus he comments on *Metaphysics* X.2, 'It is doubtful whether that which is called *one* is a substance and subsistent in itself, as the Pythagoreans said, and later, following them, the Platonists' (*com. in met.* 1962).

Consequently the question whether *ens* and *unum* are really substances, or what else they are, becomes one of the most fundamental problems of metaphysics. The problem is set in these terms by Aristotle himself in the third book of the *Metaphysics*, where he outlines the main problems of philosophy, and Aquinas consequently treats it in the same way in his commentary.

> Among all the other questions that arise, one is more difficult to consider, . . . in which it is also most necessary to know the truth, whether being and one are substances of things . . . or on the contrary . . . what it is for which it is appropriate to be *one* or *being* (*com. in met.* 488).

Plato and the Pythagoreans teach that *ens* and *unum* are not attributes of any being with a determinate content, and maintain that *unum* and *ens* are the nature of things themselves, as if being and unity were the substance of things themselves (489). Against this definition of unity as substance, Aristotle and with him Aquinas object that it is excluded by the fact that unity has the character of a universal, and a universal can never be a substance (*Met.* VII.16.1040b23 and X.2.1053b16. *Com. in met.* 1637–40, 1963–6).

The question whether the Pythagoreans and Plato really taught this doctrine is one of secondary importance. What is important primarily is the purely systematic problem as to whether it is possible to interpret unity as a substance. There is no doubt that Plato did not take up this position in all the dialogues, and especially not in the *Parmenides* and *Sophist*, but it is equally certain that in other dialogues Plato at least expressed himself in this way, and it is at all events noteworthy that Plotinus explicitly opposes the Aristotelian thesis and attributes independent being to unity. The primary intention of Aquinas's historical account is to outline the systematic problem.

Following the first possibility, namely that of regarding unity as substance, there is a second possibility, that of regarding unity as an accident. The passage from the *Summa* quoted above continues, 'Avicenna, on the contrary, . . . thought that the one which is convertible with being added something to the substance of being, as white to man' (*s.th.* I,11,1). This report of Avicenna's teaching can be supported by Avicenna's own words.[2] Avicenna actually regards unity as something like being white and thus as an accident.

Thus Avicenna also opposes Aristotle explicitly, for Aristotle points out again and again that unity is not to be regarded as an accident. It leads, as Aquinas shows, working out the idea of unity as an accident, to an infinite regress, for if unity is itself an accident, a further unity is immediately necessary to bind this accident into a unity with its substance, and so on to infinity.

It is quite impossible to define unity either as substance or as accident; it is in principle something different and transcends all these determinations and divisions of being. We may leave aside for further investigation the question whether Plato reached an understanding of the special character of the concepts of being and unity in the so-called dialectical dialogues, the *Parmenides* and the *Sophist*. But in any case for Aristotle their special character, that of transcending all articulations and divisions of being, is really his starting-point, and this is brought out particularly clearly by Aquinas in his commentary. '. . . *one* accompanies all other predicaments and is not in any one predicament alone, neither in substance nor in quantity, nor in any other, and the same is the case with being' (*com. in met.* 1975). Whatever kind of articulation or division may be made, whether into categories or in any other way, every member on either side of the division is one and has being, and the concepts of one and of being therefore transcend all possible articulations. On these grounds unity is called a *transcendens*. It is maintained by Prantl and others that the term *transcendens* only occurs in a work spuriously attributed to Aquinas. That is not true; the term occurs several times, but it is correct to say that Aquinas usually uses other terms and that the terminology of transcendentals was first fixed by Duns Scotus.[3]

We thus find as the first character of unity its transcendental character. Unity is a transcendental concept because it transcends all articulations and divisions of being.

It follows immediately from the transcendental character of *ens* and *unum* that these two transcendental concepts are convertible. Generally speaking, all transcendental concepts—Aquinas enumerates being, one, true, good, thing, (*res*) and something (*aliquid*)—are convertible with one another. Convertibility is developed early in the *Summa*, in Q. 5, Of Goodness in General. In Q. 5, Art. 3, Aquinas discusses the question, whether every being is good (*utrum omne ens sit bonum*), and gives the definition 'I answer that, Every being, as being, is good'. Here again the convertibility of being and one, which is our theme, may well be the most important point.

Whether this inner connection between being and unity underlies the Platonic dialectic could probably only be made clear in a special investigation. Aristotle, at all events, teaches the thoroughgoing connection of being and unity as part of his doctrine. Aquinas brings out the problem particularly clearly in his commentary on the *Metaphysics*. As a matter of terminology alone, Aquinas likes to speak of the one which is convertible with being (*unum, quod convertitur cum ente*), as in the passage from the *Summa* which we have quoted (*com. in met.* 1980). In his commentary he gives an account of the problem when he is discussing the relevant chapters of the *Metaphysics*, in particular Book IV.11 and Book X.3.

Of this ancient and important connection between *ens* and *unum*, all that survived was the empty Scholastic formula, *ens et unum convertuntur*. In the second edition of the *Critique*, Kant alludes to this proposition in the notes at the end of the table of the categories, not without irony.

> In the transcendental philosophy of the ancients there is included yet another chapter containing pure concepts of the understanding. . . . They are expounded in the proposition, so famous among the Schoolmen, *quodlibet ens est unum, verum, bonum*. Now although the application of this principle has proved very meagre in consequences (which yielded nothing but tautological propositions) and therefore in recent times has retained its place in metaphysics almost by courtesy only. . . . (B 113).[4]

But the significance of the connection is very far-reaching, and one can in fact say that in Kant, as we shall see, the being of appearance, the primary definition of which is for Kant an ontological definition, rests on the unity of transcendental apperception, and that for Kant as for Aquinas, unity and being (of possible experiences) are convertible.

The common transcendental character and the convertibility of *ens* and *unum* are still in a sense categorial definitions, but they lead immediately to the first ontological definition which is the subject of Q. 11, Art. 1 and is expressed in the thesis, 'I answer that, One does not add anything to being, but only the negation of division.' Here again the *Summa Theologica* gives the conclusion, which is crystal clear, reached in the commentary on the *Metaphysics*. The decisive passages are *Metaphysics*, IV.2 and X.3 with the corresponding passages of the commentary. In the commentary on IV.2 Aquinas says, 'The one which is convertible with being . . . does not suppose any nature added to being' (*com. in met.* 560). In *Metaphysics* X Aristotle himself had already clearly outlined the matter by saying that 'one man' expresses no more than 'man', so that unity does not express any further content of being beyond being (*Met.* X.2.1054a16). In commenting on this passage Aquinas brings out the point very clearly (*com. in met.* 1976), but perhaps the clearest expression of it is in the explanations of the subject at the end of the commentary.

> He shows a little later that they signify the nature of those things of which they are said, and not anything added, such as an accident. For in this, *communia* differ from accidents, and however much is common to the two is of no importance; since *communia* mean the nature itself of the things of which they are predicated, but accidents mean some added nature (1980).

By *communia* here is meant the transcendentals, in particular *ens* and *unum*. They thus have to be essentially distinguished not only from substances but also from accidents, because they signify being in itself, and not anything additional. The wrath of Achilles signifies something which can be added and can be taken away again, but the unity of Achilles signifies Achilles himself in a completely irreducible way. The transcendental determinations refer to what is in itself, and they thus become the fundamental determinations of being.

But the question at once arises, if the transcendentals denote nothing additional, do they denote anything at all? Are they not completely empty? Aquinas sets himself the problem in the third objection to Q. 11, Art. 1. If the expression 'one man' says nothing more than the expression 'man', then to predicate unity of an existent can say nothing new. *Unum* and *ens* would then be the same, so that one would be saying in a completely empty way, *ens ens*. And thus all transcendental determinations would be nothing but completely empty and meaningless expressions. Now the transcendental determinations can be said of substances and of accidents, I can say that a man is and that he is one, and I can say equally that a colour is and that it is one. The question thus arises, in what way are the transcendental determinations different from each other and from the determinate content of the substances and the attributes of which they are predicated. In what way do the being and the unity, which I predicate of this man here, differ from the man himself, and in what way do they differ from each other? In what way are the being and the unity which I predicate of this colour different from the colour and from each other? The problem is particularly acute for Aquinas because he is discussing the being of God. I can predicate unity and goodness of God, but I still have to ask, in what way are these predicates different from God himself and in what way are the goodness of God and the unity of God different from each other? Aquinas's answer is that the transcendental determinations do not differ from the substrates 'really' (*secundum rem*) but only 'in idea' (*secundum rationem*). This is true both of the divine being and of created being, and both of the good and the one. We may quote Q. 5, Art. 1, where Aquinas says, 'I answer that, Goodness and being are the same really (*secundum rem*) and differ only in idea (*secundum rationem*).'

This means therefore that both in divine and in created being unity and perfection do not differ 'really' but only in respect of *ratio*. This touches the fundamental problem, that in the being of unity the *ratio*, that is the understanding, is in some way involved. For Plato the understanding was involved in the being of ideas, if we may follow Natorp's interpretation at this point; but the problem in Plato may wait for further investigation. Aristotle, as we know, brought out the connection between unity and *logos* explicitly (*Met.* IV.2.1003b22).

We therefore also find Aquinas's definitive account of the problem in his commentary on this passage. He asserts the proposition, 'One and being signify one nature according to different reasons (*rationes*)' (*com. in met.* 549). He then gives a full account of the way in which being and one are identical 'really', and in what way and on what grounds they are different according to *ratio*. He summarises his argument as follows.

> It is therefore clear for the reason that has been given not only that they are one really (*re*) but also that they differ as to reason (*ratione*). For if they did not differ as to reason they would be entirely synonymous, and it would thus be indifferent whether one said *ens homo* or *unus homo* . . . hence these three, *res*, *ens*, *unum*, mean exactly the same, but according to different reasons (*rationes*) (553).

This reference of the transcendental determinations to *ratio* is brought out very clearly and very carefully in the *Summa*, in Q. 5, Art. 1, where the good, the one, and the true are discussed. It leads to an important consequence which we wish to consider in the case of the one. What does it mean when we say that God is one? The reference of the one to *ratio* involves a privation in the concept of one, and unity is in fact defined by Aquinas as undividedness. But then if we say that God is one, we are predicating a privation of God, and this is impossible: 'Likewise neither can *one* which is convertible with *being* be predicated of God, because it imports privation, and every privation is an imperfection, which cannot apply to God. Therefore it cannot be said that God is one.' Aquinas's reply to this objection is that when we know God as the one, we only know him in the mode of our apprehension.

> And although in God there is no privation, still, according to the mode of our apprehension, he is known to us only by way of privation and removal. Thus there is no reason why a certain kind of privation should not be predicated of God, for instance that he is incorporeal and infinite, and in the same way it is said of God that he is one (*s.th.* I, 11,3).

The reference of unity to *ratio* thus brings with it the reference of unity to our *ratio* and it follows that our representation of the unity of God is only a representation after the manner of our possible ways of knowing.

With this reference of unity to *logos* Aristotle reached one of the most fundamental parts of metaphysics. That Aquinas takes up this standpoint with such energy and such clarity shows that in this point, as in others, Aquinas's acceptance of Aristotelian metaphysics represents a genuine and deep understanding of it. The boldness with which Aquinas uses the transcendental philosophy of Aristotle for the problems of Christian dogmatics is astonishing. One cannot escape the impression that it was precisely the ontological problems of dogmatics which were the decisive motive for the return to Aristotelian ontology. It would be unreasonable to complain that Aquinas's account is still encumbered with many difficulties. The reference of being, unity, perfection, and truth to *logos* belongs to the central problems of ontology. Anyone who has ever tried to think out these problems for himself and has seen how great were the difficulties with which Plato, Aristotle, Leibniz, and Kant had to struggle when they dealt with the same problems, will not complain of the difficulties in Aquinas, but will admire his achievement. It must not of course be overlooked that many pupils of Aquinas failed by far to reach the depth, the mobility, and the quality of genius in the master's thought, so that many strictly Thomist accounts have little to say about Aquinas's deepest insights. Though it should perhaps not be forgotten that it is always the case—and certainly among Kantians—that pupils fall behind the master. It may be wondered at, that in this interpretation of the *Critique of Pure Reason* so extensive an account of Aquinas is attempted. It is true that we depart as much from the opinion of orthodox Kantians as from that of orthodox Thomists. An orthodox Kantian is as much a contradiction in terms as an orthodox Thomist. We believe that philosophy is always one, in its historical links as in its systematic connections, and we believe that Plato and Aristotle, Augustin and Aquinas, Leibniz and Kant, were struggling with the same great questions, what is unity? what is truth? what is being? With the reference of unity to *ratio*, which Aquinas is able to clarify further by continuing the Aristotelian ontology, we see the genuineness of the primordial system of connections in the *philosophia perennis.*

There remain two special problems in Aquinas to be briefly mentioned because they were important for later developments. They are the problems of transcendental number and transcendental relation. We saw that Aquinas distinguishes between transcendental and categorial unity, between the one which is convertible with being and the one which is the principle of number. Categorial unity adds a new content to its subject. It is bound up with extension, with *quantitas continua.* The same considerations hold for numbers, which arise from categorial unity as their principle. Number, like unity, is a categorial determination bound up with extension. It follows from this that substances which are not extended cannot participate in categorial number, and the trinity of divine persons or a trinity of angels cannot therefore be a categorial number. Aquinas solves this problem by developing the distinction between the two modes of unity, and distinguishing a transcendental and a categorial number. Although Aquinas, so far as I can see, only speaks of transcendental plurality, I cannot see any objection, in view of the circumstances, to following the majority of commentators, for instance Suarez, Gredt, and Bodewig[5], and speaking of a transcendental number in Aquinas. The problem of transcendental relation is rather more difficult. Aquinas's theory of relations is one of the most difficult parts of his philosophy, and the question whether Aquinas really distinguished between real, transcendental, and rational relations is not easy to answer. I follow Duns Scotus, Ockham, and Suarez, and say that the doctrine of transcendental relations is present at least implicitly in Aquinas's writings.

In returning to Aristotle and thereby to the Aristotelian ontology, in particular the problem of transcendentals, the Scholastics made their fundamental ontological decision. Aquinas carried out this movement and in so doing, raised Scholastic philosophy to an eminence which was not maintained and was never subsequently reached in Scholasticism. That is true even of the two great thinkers of late Scholasticism, Duns Scotus and William of Ockham. I gave a detailed account of the problems in my book *W. v. Ockham* (1949). For our present more extended enquiry we may be satisfied with a short summary. The enquiries of these two thinkers start from Aquinas's distinction between transcendental and categorial unity. It had already been shown by Aquinas that this distinction could not remain confined to unity. Aquinas himself made the same distinction between transcendental and categorial number, and at least prepared the way for a similar distinction between relations. Duns Scotus follows up the distinction. It is not easy to say whether he reaches definitive conclusions as regards the problem of number or whether he merely raises the problem for discussion. But as regards relations, he distinguishes in full detail between real, rational, and transcendental relations. He thus retains Aquinas's distinction and even extends it. Ockham, starting from the same point, reaches quite a different conclusion. He too starts from the distinctions between transcendental and categorial unity and number in Aquinas, and between transcendental and real relation in Duns Scotus. But he points out that both modes of unity, of number, and of relation are in an extended sense real and therefore raises the question whether it is necessary to make this distinction at all, and whether one cannot make do with allowing reality to unity, number, and relation in the sense of the transcendental modes. If the number of angels, for instance, can only be a transcendental number, the question must arise as to whether this transcendental number alone is not enough, so that every number would be interpreted as a transcendental number, a number of men just as a number of angels. Ockham then investigates all the categories in the same way. He comes to the conclusion that all the categories, with the exception of substance and quality, are to be interpreted as having transcendental being. All categories, with the exception of substance and quality, have the same transcendental being as transcendental unity, transcendental number, and transcendental relation. Ockham thus has to deny the reality of all the categories except the first two, but this denial of the reality of quantity, relation, etc., has exactly the same meaning as Aquinas's similar thesis in respect of the reality of transcendental unity. In the same sense, and only in the same sense, in which Aquinas denies that transcendental unity is something added, *res addita*, does Ockham deny that quantity and relation in general are to be regarded as something added. Ockham's denial of the reality of quantity and relation is therefore not nominalism but transcendental philosophy. From this point of view Duns Scotus's and Ockham's further ontological treatment of the distinction between categorial and transcendental being can be called a continuation and development of Aquinas's thought.

This would then bear on the question of the extent of what is to be called transcendental being. Aquinas carried the idea of transcendental being into the categories, Duns Scotus continued the same train of thought, and Ockham brought it to a radical conclusion by interpreting all the categories except substance and quality as having transcendental being.

On the fundamental question of how transcendental being itself is to be understood, the two later thinkers probably did not entirely reach the eminence which Aquinas reached. We found that Aquinas's fundamental conclusion was his insight into the reference of transcendental being to *ratio*, in particular the reference of unity to *ratio*. Duns Scotus takes up the subject of what kind of being the transcendental determinations may be said to have, and constructs for the purpose the new concept of formal being. The transcendental determinations have formal being. Now the problem of formal being is one of the most difficult problems in the philosophy of this subtle thinker, but it seems that the reference of formal being to *ratio* and therewith the reference of the transcendental determinations to *ratio* is forced well into the background. If this is so, the boldness with which Aquinas made Aristotle's—and perhaps also Plato's—connection between unity and *logos* into his fundamental thesis was no longer fully effective with Duns Scotus.

One would expect that Ockham, on the other hand, who is so readily called a nominalist, would have developed this reference of the transcendental determinations to the *logos* into a consistent nominalism, and that the being of transcendental determinations, and in particular of unity, would consist for him only in their being thought. But my study of Ockham has shown that this is by no means so, and that on this fundamental question of the ontological status of transcendentals, he discusses a whole series of possibilities but decides for none of them. It is noteworthy that Ockham does not seriously consider the purely nominalistic solution and does not even regard it as a discussable possibility. One therefore has to say that on this fundamental ontological question Ockham did not reach the eminence and the boldness of Aquinas, and that his importance lies in his consistent carrying through of the transcendental viewpoint. Ockham's philosophical writings therefore have as it were a quantitative importance. They refer primarily not to the question, what is transcendental being?, but to the question, how far does transcendental being extend? Here Ockham's importance is such that we shall give a separate account of the problems in the next section.

§ 17

Universals and Relations

It is not unusual among friends of mediaeval philosophy to approach the problem of generality from the problem of universals. If this point of view is adopted, problems of unity are reduced to those of the unity of a genus. Now not only was the problem of universals exhausted, through endless discussions in which the problem of the unity of a genus was turned every possible way; it also came about that the new natural science founded on mechanics had no further use for the concept of a genus. But if we turn our glance backwards we see that neither in Greek nor in mediaeval philosophy is the concept of a genus the only general concept, and it may not even be the most important one. Aristotle declares explicitly that the concepts of unity and being, although they are general concepts, are not generic concepts (*Met.* II.3.998b22). Aquinas repeats that all transcendental concepts are general concepts but they are not generic concepts (*com. in met.* 433, *s.th.* I,3,5). The fundamental concepts of ontology are not universals, or at least not in the sense in which the problem of universals is almost always discussed. For Aquinas there is the additional factor that he is primarily concerned with the ontological questions of theology, and as generic concepts play no part in this problem, he is led to consider almost exclusively the transcendental determinations of God's being, namely unity, perfection, and truth. From Aquinas's point of view it must therefore seem very questionable to restrict oneself to generic concepts as the only general concepts. For Kant mechanics has come into the foreground. The Kantian philosophy can only be understood if one keeps in mind that the problems of unity which Kant considers have no bearing on the unity of a genus. On the systematic side we are encouraged in this interpretation by the fact that Husserl has at last made the different possibilities of the general clear again in his distinction between generalisation and formalisation.[1]

The generality with which we are concerned in mathematics, physics, and hence in the *Critique of Pure Reason*, is that of number and of relation. It was Ockham who first grasped these problems systematically, and here again, as so often, the problems appeared in a particularly clear light the first time they were grasped. In a certain sense Duns Scotus can be said to have seen the problem when he raised the question as to the being of numbers; what he actually discussed was the being of the natural numbers 2, 3, 4 . . ., in accordance with the Greek conception. He does not seem to have come to a definitive conclusion.[2] But he reached the important result that at all events the concept of a genus cannot fit the relation of the particular number to the general concept of number—as we should say today, the general concept of natural number. Number is not the genus of 2, 3, 4, . . . If it were, the particular numbers would have to be considered as species, and the transition from genus to species would have to take place through a specific differentiation. But as the concept number would have to have an infinite number of species, there would obviously also have to be an infinite number of specific differentiae, and as Duns Scotus correctly says, this is impossible in the logical system of genus and species. The correctness of this view is supported by the fact that Plato encountered the same difficulty. The attempt was made in the Academy—whether by Plato himself or his pupils we cannot exactly tell—to subsume numbers under the dichotomic schema of genus and species, but this attempt met with no success, obviously because the schema of genus and species is quite unsuitable.

The same problem arises again in a more acute form in the relation of number to that which is numbered, and Ockham seems to have been the first to see this. The particular number is itself general. This is especially clear in the case of the larger numbers. The concept seventeen, when it is applied to seventeen men, is something general, and the question is, what is the nature of this generality.

The problem of the ontological determination of number is first dealt with by the Scholastics by means of the concept of accidence. Number is regarded as a categorial determination, as a sub-species of quantity, and the ontological status of everything categorial is fixed by the strictly accidental character of qualities. The number two is first regarded as an accident in this sense. But Aquinas had to posit transcendental plurality as well as accidental number, and Duns Scotus and Ockham showed definitively the difficulties which arise if one tries to regard number as an accident. Take for instance the number two, as in two concrete things. What is the ground—this is the question—of the fact that these two stones are two? If we try to conceive the twoness of these two stones as an accident in the proper sense, we soon find ourselves in difficulties from which there is no way out. Either the accident of being two is wholly in each of the two stones, and that is obviously absurd because one of the two stones could then disappear and the other would still have the quality of twoness. Or being two is a quality that exists somehow between the two stones, with one foot in one stone and one foot in the other. But this is in contradiction with the concept of an accident. Or finally one part of the twoness is in one stone and the other part in the other stone, but the question would then only have been removed a stage backwards. We should now have to ask what gives the two parts the quality of being two, what makes them a two, and we should be back at the original question again.[3] It comes to light in this discussion that in the two, unity and plurality are intertwined in a highly characteristic way. Two is in a quite specific way unity and plurality at the same time. The question in which we are primarily interested is the question of unity. What is it that makes these two stones represent a new unity in their twoness, even if it is a quite peculiar unity?

There is a very old and very simple answer. These two stones are only two because they are being considered by an understanding, generally a man, which is counting them. This reduction of number to the man enumerating is something that one would expect to find in the nominalists generally, but it does not occur in Ockham and is rare in Scholasticism in general; as far as I know, the only important Scholastic to have held this view is Petrus Aureolus. But none the less Aristotle's remarks about the connection between number and the enumerating soul made a deep impression on the Scholastics, not least on Aquinas. Aristotle had simply said, 'If there cannot be someone to count, there cannot be anything that can be counted, so that evidently there cannot be number' (*Physics*, IV.14.223a22). It was only very slowly that philosophers dared to take this proposition of Aristotle at its face value, although Plato had drawn attention in the *Timaeus* (37d) to the connection between time and number.

The problems show up even more clearly over the question of relations, and our present-day insight into the relational character of number has made this connection between number and relation intelligible. Over the problem of relations Scholastic ontological discussion again has to struggle with the concept of accidence.

Relation is first included among the categories and being thus associated with quality, is treated ontologically in the same way, namely as a strict accident. The equality of the two pieces of wood is then an accident. But again, as with the two stones, the question arises, how and where does the equality exist? Is it wholly in each of the two, or floating between them and in neither of them, or is it partly in one and partly in the other? All the answers are impossible and they show the impossibility in principle of this approach. Equality cannot be called the property of a thing. Equality and similarity always characterise two things, and Ockham found the most concise expression for this when he said: *similitudo sunt duo similia.*[4] Here it is seen as clearly as possible that a relation binds two things together into a unity, and there arises the general question, on what does the unity of the relation rest?

Again we have the old and too simple answer, that relation consists in the relating or comparing of two things with one another by an understanding. Two things are equal because an understanding compares the two things with one another and represents them to itself as equal. This thesis is favoured by the fact that there certainly are relations which rest solely on their representation by an understanding. Such relations have always played a great part in Scholasticism under the name of *relationes rationis.* But it cannot be possible to ground all relations solely on their being represented and to regard all relations as *relationes rationis.* And if real relations in the proper sense are admitted, the question arises, in what does the reality of the real relations consist?, and this then means, in what does the unity of real relations consist?

§ 18

Unity in Leibniz's Ontology

Leibniz is equally important both historically and systematically for our present purpose. Historically Leibniz lives in an unbroken continuity with Scholasticism, and Kant takes his problems directly from Leibniz in the form in which Leibniz posed them. Systematically it emerges as clearly as one could wish in Leibniz that there are two quite different problems, the unity of the living and the unity of the phenomenal. Leibniz's name for the unity of living things is the monad. Monad and living beings are co-extensive concepts. Every living being is a monad and every monad is a living being (if we disregard God). The concept of monad brings the problem of unity into special prominence. These problems of the unity of living things lie outside the scope of our enquiry, but from our point of view one could say that for Leibniz the essence of the living is its unity. In this enquiry we are only concerned with the unity of phenomena, which culminates in the unity of nature, nature being regarded here as a mechanical system.

The unity of this mechanical system has a thoroughly relational character for Leibniz. It is a unity of relations, a conformity of relations to law, and as such a relation of relations.

This reduction of the phenomenal to relations as a matter of principle is a continuation of what Ockham had achieved. For this summary account we may be satisfied with alluding to the fact that Leibniz subsumes quantity under relations. This is true even of numbers, and Leibniz can say in a letter to Des Bosses that 'numbers, units, fractions have the nature of relations' (G.II,304). In a fragment published by Couturat, Leibniz explains at length that he had at first carefully distinguished between quantity and relation as two different categories. But when he had worked over the problem more carefully—*re tamen accuratius considerata*—he had seen that all quantitative determinations are relations.[1] We have already indicated how decisively Leibniz emphasises that mathematics is theory of relations. For Leibniz the question, what is nature?, as well as the questions, what is space? and what is time?, are all subsumed under the more general question, what is relation? It was Ockham who first developed this general point of view. Now all real existence is co-extensive with the existence of monads. What is real is a monad, and relations cannot therefore be real in the proper sense. Here Leibniz takes up the extreme standpoint that all relations as such belong to thought. '*Relationem . . . esse rem mere mentalem*' (G.II,486). Relation, this must presumably be translated, is a pure figment of thought. There are thus only monads and in a certain sense qualities as modifications of monads, and there are no relations until they are added by thought. 'I believe that qualities are only modifications of substances and that the understanding adds to them relations' (G.V,132). This would admittedly mean an extreme nominalism if the thinking that adds the relations were not God's thinking. We have already seen the connection with God's thinking in the case of space, and we now meet it in the general form. Leibniz found the best formulation of it in a passage which we only know in draft, in the draft of a letter to Des Bosses. 'Thus God not only perceives the individual monads and all their modifications, but also their relations, and in this consists the reality of relations and truths' (G.II,438). The reality of relations consists in their being thought by God.

This reference of relations to the divine thinking is not nominalism but transcendental philosophy. In comparison with Aristotle's original approach there are two changes. It is recognised that relations are unities, and the question of unity is confined to those unities which are relations. At the same time the connection between unity and *logos*, which Aristotle had brought out, becomes a central theme. We shall try to clarify Leibniz's position by sketching in outline what are in principle the possibilities of transcendental philosophy.

By transcendental philosophy we mean that part of metaphysics which is concerned with the transcendental determinations, that is to say, in which the questions are asked, 'what is unity?' 'what is perfection?', 'what is truth?', 'what is being?' The question whether transcendental philosophy exhausts metaphysics, that is, whether all metaphysics is transcendental philosophy, may be left undecided. Further, we are determining our position solely with regard to the question 'what is unity?', and we limit the question to such unities as are relations, unless all unities conceivable by us should be relations. Three stages, or one could say three standpoints, of transcendental philosophy are then possible. The first standpoint emphasises primarily the distinction between unities and things, the second emphasises that unity has something to do with *logos*, and the third standpoint grounds unity wholly on *logos*. The Kantian transcendental philosophy is indeed a specifically transcendental philosophy, but it is unjust to regard it as exclusively one of the third stage, though the question whether it is really and exclusively such may be left undecided for the time being. The interpretations of Kant favoured in Neo-Kantianism and in German Idealism are attempts to understand all unity from thinking alone, but it is still questionable whether this interpretation really exhausts the Kantian philosophy. However that may be—we shall return to the point later—it is very important for understanding and justifying the Kantian standpoint to exhibit these three stages in their proper sequence.

The first stage is the ontological assertion that unity is something other than the things of which it is the unity. This assertion we have already met in many concrete forms. Space was primarily a system, a system of order, a unity of things, and as such it could not itself be a spatial thing. Similarly, nature as the conformity of things to law cannot itself be a thing standing under these laws of nature. The point of this first stage can be seen most clearly in the case of the laws of nature. A law of nature presents the connection between things according to law and hence cannot itself be a thing. Aristotle and following him Aquinas distinguish the special being of unities from the being of substances and of accidents. Unity is neither a substance nor a quality, as Aristotle said with all possible clarity, and we saw that Aquinas, taking up this theme, had achieved a genuine understanding of Aristotle.

The second stage is the question, in what does the being of unity consist if it can be neither the being of a substance nor the being of a quality?, but we do not find a detailed answer to this question in Aristotle. Aristotle is satisfied with indicating the importance of *logos*. It is in *logos* that the differences between the transcendental determinations are grounded. If we have understood Aquinas correctly, this reference of unity to *logos* or *ratio* is brought out more sharply by Aquinas, presumably in order to deal with the ontological problems of fundamental theology. The difference between transcendental determinations is grounded in *ratio*—this is the fundamental thesis of Aquinas's transcendental philosophy.

From this point a third stage is possible, if the unity of relations (with which alone we are now concerned) is not only referred to *ratio* but wholly grounded in *ratio*. The being of unity is not then merely referred to thinking but consists, for this third position in transcendental philosophy, solely in its being thought. This viewpoint perhaps found its most mature expression with Cohen, in his systematic philosophy as well as in his interpretation of Kant. For him all unity is thinking, all unity is grounded wholly in thinking. The being of space, of the order in nature, of unity in every form, consists exclusively in its being thought.

A satisfactory understanding of the Kantian philosophy is only possible if the Kantian standpoint is consciously related to these different possible forms of transcendental philosophy. There is a further distinction which originates from the fact that the *ratio* can be either that of God or that of men. If the *ratio* of God is admitted as the *ratio* in which all existence is grounded, then all phenomenal unity, the unity of all quantities, all numbers, and all relations, consists for Leibniz in their being thought, though of course in their being thought by God.

§ 19

The Transcendental Ideality of Unity

To approach Kant's final answer to the question of the ontological status of unity from Leibniz's position again, there is the underlying distinction between substantial unity and phenomenal unity. Substantial unity is the unity of the living and is expressed in the monad, phenomenal unity is the unity of natural appearances in so far as it is studied in the natural sciences founded on mathematics. This unity originates in thinking, and the thinking in which it originates is God's thinking.

Kant retains this distinction and goes further, in that the starting-point for the discussion of unity in the Transcendental Analytic is the unity of natural appearances alone. He also retains the view that the problems of phenomenal unity are problems of relations. For Kant, as for Leibniz, space was a system of relations and mathematics was theory of relations. Nature consists exclusively of relations, nature is a pure problem of relations, and mathematico-physical natural science is a theory of relations. The turn which Kant gives to the problem is that relations are now considered exclusively from the point of view of unity. Relations, as Ockham had shown, are unities, and they are now understood by Kant wholly from the point of view of their unity. This is perhaps most clearly to be seen in the case of the causal relation. The causal relation is a relation, but its essential significance lies in the fact that it establishes a unity, in the first instance a unity between two things, of which the one is cause and the other effect, and then the specific unity of nature, since the causal relation pervades nature as a whole; nature is one because it is bound together in a system of causal connections. All the unities which are investigated in the *Critique of Pure Reason* are relations.

But none of these relations are given in things; they all rest on the production of these connections in thinking. A relation between two things consists solely in the referring of the two things to each other in thinking.

Kant develops this idea in the Transcendental Deduction. 'Thus the order and regularity in the appearances which we call nature, we ourselves introduce, and we could never find them there had not we ourselves, or the nature of our mind, originally set them there' (A 125). This sentence from the first edition Transcendental Deduction concentrates the problem as in a burning-glass. The concept of nature is first expounded once more. Nature is not things and not even living beings as such; nature, as we have already indicated, is solely order, regularity, conformity to law as such. This order and regularity consists in a connection of appearances such that appearances are subjected to a thoroughgoing unity. Connection, and hence the unity produced by the connection, do not lie in the things themselves but are put into them by us. This connecting is not an occasional and accidental activity, but the essential feature of our thinking. Thinking means connecting into order and regularity. Kant sees fully the significance of describing thinking in this way, for he says in the next paragraph of the Transcendental Deduction:

> We have already given various different explanations of the understanding; as a spontaneity of knowledge, . . . as a faculty of thinking, or as a faculty of concepts, or again of judgments . . . We may now characterise it as the faculty of rules. This distinguishing mark is more fruitful and comes closer to its essential nature (A 126).

The understanding is thus now characterised in a more fruitful and more essential way as the faculty of rules. 'Understanding' means 'the faculty of rules'. But all these rules, since nature consists wholly of relations, are relations, and the understanding is thus defined as the faculty of positing relations. Putting relations into things is no occasional and accidental activity of the understanding, but constitutes the peculiar essence of the understanding.

We have so far kept to the account in the first edition. The second edition in no way takes back what has been said but rather intensifies it.

The Transcendental Deduction in the second edition begins with this same thesis.

> But the combination (*conjunctio*) of a manifold in general can never come to us through the sense, and cannot also, therefore, be contained in the pure form of sensible intuition. For it is an act of spontaneity of the faculty of representation; and since this faculty, to distinguish it from sensibility, must be entitled understanding, all combination—be we conscious of it or not, be it a combination of the manifold of intuition, sensible or non-sensible or of various concepts—is an act of the understanding. To this act the general title 'synthesis' may be assigned, as indicating that we cannot represent to ourselves anything as combined in the object which we have not ourselves

previously combined, and that of all representations *combination* is the only one which cannot be given through objects. Being an act of the self-activity of the subject, it cannot be executed save by the subject itself (B 129).

Kant uses the term *Verbindung* (combination) here, because in the German language this term can express both a state and an act and thus makes it particularly easy for him to go back from the state to the act. There is only a combination where we ourselves have combined something. There is only a relation where we ourselves have related things to each other. This combining is now emphasised with renewed energy as the essential character of the understanding.

Combination does not lie in objects and cannot be borrowed from them through perception for instance, and only then taken up into the understanding. On the contrary it is an activity of the understanding alone, which itself is nothing but the faculty of combining a priori (B 134).

This definition of the understanding as the faculty of combining a priori is the crimson thread of the Transcendental Deduction in the second edition.

The problems that are brought together here at the focus of the *Critique of Pure Reason* have always been carefully brought out in the great interpretations of Kant. Cohen, Heidegger, Hartmann, Heimsoeth have all emphasised this central problem from their different points of view.

In his interpretation of the *Critique of Pure Reason* Cohen considers the *Critique* as a theory of Newtonian physics. This attitude is not so one-sided as for a long time may have appeared. The weaknesses of Cohen's interpretation lie, as we shall see later, in the problem of the thing in itself, but in the problem of unity Cohen reaches the real kernel of Kantian thought. Cohen formulates the problem by using the term 'unification'. All unity is unification. 'Synthesis has to produce unity, it is therefore a unification.' The act of unifying determines in principle the nature of thinking. Thinking, says Cohen, is unifying; 'Every kind of thought is characterised by this task, they all aim at unification of the elements of consciousness which are offered in the manifold of intuition, however specialised may be the kind of unifying performed by each kind of thinking' (p. 326).[1]

Heidegger, in his interpretation[2] of the *Critique of Pure Reason*, disregards all problems in theory of science and concentrates on the problems of metaphysics. But from this point of view again the categories, the modes of unity, are interpreted essentially as modes of unification. 'Unity is fundamentally unifying. That is to say: the representation of unity takes place as a unifying, for which, as a structural whole, the intention of unity is required' (§ 17a). The manifold of the categories thus becomes a manifold of modes of unification, 'as a complete whole, a manifold of modes of unification' (§ 16). Establishing unity, unifying, finally becomes the original act of the understanding. If Cohen and Heidegger agree thus far, this may be an encouragement to us to seek the central problem of the *Critique* in the conception of unity as unification.

Kant has now reached a twofold conclusion. The categories are ways of unifying, and unifying is not an accidental occupation of the understanding but is the understanding itself. The general Kantian term synthesis now gains a fuller significance. Synthesis now means the spontaneous, constructive, active moment. If it was undecided until now whether synthesis meant the state or the act, its meaning is now decided in favour of the act. Synthesis now means the real bringing together of different things, and synthesis, which is this bringing together, is regarded as the essential characteristic of thinking. From this thesis, that all combination consists in the understanding combining, that all relation consists in the establishing of relations by the understanding, the Transcendental Deduction goes forward to a new and equally fundamental proposition, to establish which is its real aim. The understanding was called the faculty of combining, the faculty which creates ordering and regulating relations. Nature, for its part, is nothing but a system of relations, of combinations. Thus understanding and nature belong together in a primordial way. The understanding is the faculty of positing relations, nature is a totality of relations. There is an original connection between nature and the understanding, and nature is nothing but the coming into force of the original potentiality of the understanding. The connection is established by reducing both nature and the understanding to relations. Nature is nothing but a system of relations and the understanding nothing but the faculty of establishing relations, and nature must therefore arise out of the activity of such an understanding. Nature and the understanding are conjoined in this relational character, and Kant can therefore ask,

But how could we bring about a synthetic unity a priori, if subjective grounds of such a unity were not contained a priori in the original sources of knowledge in our mind and if these subjective conditions were not at the same time objectively valid, being the grounds of the possibility of knowing any object at all in experience? (A 125).

And further:

> The a priori conditions of a possible experience in general are at the same time conditions of the possibility of the objects of experience (A 111).

The conditions are identical because thinking and the objects of thought are simultaneously relational in character, because the understanding and nature are equally relational in character. Nicolai Hartmann drew attention to this identity-theory, which is present here and carries the Kantian argument as a whole, in his article on the principles of interpreting Kant.[3]

These arguments which are developed in the Transcendental Deduction have been taken again and again as a hymn to the potentialities of thinking. All unities flow from thinking, all unities are unification, all being is unity and thus unification. However much these Kantian arguments are in fact a hymn to the potentialities of reason, they are at the same time—and this was completely overlooked by Cohen—an elegy on the bounds and limitations of human reason. All unities, all relations, originate in human thought, and thinking thus far has a limitless spontaneity, but all positing in thought is never more than a positing of relations. Heimsoeth forcefully drew attention to this limitation.[4] All human knowing grasps not things as such, and in particular not the essence of things as such, but only relations between things. The same factor which is the ground of the spontaneity of thinking, because all thinking is concerned with relations, is at the same time a limitation of thought, because all thought is only concerned with relations. On the one side we have in Kant's philosophy the pride and the optimism of the Enlightenment; it is the understanding which prescribes laws to nature. But all these laws which the understanding prescribes are nothing but relations, they are the ground only of a complex of relations, they are the ground only of appearances.

The unity whose being is under examination in the *Critique of Pure Reason* flows entirely and in its whole content from thinking, but the unity generated in thinking is never more than the unity of relations, and thus the unity of appearances, and is not being in itself. The decisive thesis of the *Critique of Pure Reason*, that thinking unifies all unity, is at the same time a hymn to the creative powers of reason and an elegy on its limitations.

PART TWO

BEING

CHAPTER V

THE MODES OF BEING

§ 20

The Three Great Themes of Metaphysics

ONE fundamental thesis runs all through Kant's works: the final ends of metaphysics and of human reason as a whole are the three great themes of God, freedom, and immortality. These are the three questions which human nature can never cease to ask. In a note added in the second edition of the *Critique of Pure Reason*, Kant says this with all possible clarity. 'Metaphysics has as the proper object of its enquiries three ideas only: God, freedom, and immortality' (B 395). The Preface to the second edition leads up to the same point: the unavoidable tasks of pure reason itself are God, freedom, and immortality (B xxx).

It is not true to say that in the *Critique of Pure Reason* the attitude to these problems is purely negative. The number of the fundamental problems of metaphysics, three, is first deduced systematically. In the section called System of the Transcendental Ideas, Kant shows that all transcendental ideas can be brought under these three heads. In the same way, the three rational transcendental disciplines, rational psychology, rational cosmology, and transcendental theology (A 334, B 391), arose as 'a pure and genuine product or problem of pure reason' (A 335, B 392). The *Critique of Pure Reason* then goes on, notwithstanding all the necessary criticism, to complete the first step towards a positive solution. It shows that these three ideas are free from contradiction, and that no proof directed against them can ever succeed on a purely theoretical basis. The existence of God and the facts of immortality and of freedom cannot be proved from a theoretical standpoint, but they also cannot be disproved.

The positive proof can be given on a practical basis, as Kant declares in the Preface to the *Critique of Practical Reason.*

> The concept of freedom . . . is the keystone of the whole structure of a system of pure and even speculative reason, and all other concepts (those of God and immortality) which, being merely ideas, are without foothold in

pure reason, are now joined to it, and with it and through it receive content and objective reality (V,3; Abb. 88).

Thus the three great ideas of metaphysics, the ideas of God, freedom, and immortality, can have objective reality demonstrated of them from practical presuppositions.

Here we run into one of the most difficult problems in understanding Kant. Why is it that, in the *Critique of Pure Reason*, a negative attitude is taken to these metaphysical tasks, while in the later works a positive solution is reached? In the Preface to the *Critique of Practical Reason* Kant himself calls this the riddle of critical philosophy.

Not until now do we have the explanation of the riddle of critical philosophy, how the supersensible use of the categories can be denied objective reality in speculation, and allowed objective reality in respect of the objects of pure practical reason (V,5; Abb. 90).

Friedrich Paulsen,[1] in 1898, was the first to attack this problem. He puts these three great themes in the centre of his interpretation of Kant. The starting point now becomes the distinction between two worlds, the intelligible world and the sensible world. The proof of the objective reality of God, freedom, and immortality is thereby moved into the centre of Kantian thought. But, as I believe, Paulsen misses the real riddle of the *Critique*, in that he misses the question, how the attitude of the *Critique of Pure Reason* can be reconciled with the attitude of the later works (and hence also of the pre-critical works). The reconciliation of these two attitudes is an essential problem and Paulsen does not see it, through looking at everything from the point of view of the later works and the pre-critical works. This goes so far that the *Inaugural Dissertation* becomes for him Kant's central work, while the *Critique of Pure Reason* moves to the periphery. Paulsen says explicitly in his Introduction:

In the *Critique of Pure Reason* the negative side, the attack on the false foundation, is in the foreground, and here Kantian thought is at its greatest distance from its centre. In the subsequent writings, and especially in the two subsequent critiques, the 'intelligible world', whose reality remains the presupposition that is taken for granted in the *Critique of Pure Reason*, comes to the fore in the most pronounced way as the dominating central point (p. xx).[1]

To say that in the *Critique of Pure Reason* Kantian thought reached its greatest distance from its centre sounds strange; to speak bluntly, an interpretation which reaches such a conclusion must be false from the outset, whatever grounds may be brought forward in support of it. The *Critique of Pure Reason* is the central point of Kantian philosophy, of this there can be no doubt, either from outer or inner, biographical or systematic standpoints. It is impossible to see the centre of Kantian thought in the pre-critical writings which were explicitly and rightly repudiated by Kant. While fully recognising Paulsen's achievement in being the first to take up these metaphysical problems again, it is hardly possible to suppress this criticism.

Similar criticism can be made against Max Wundt's extensive account of Kant as a metaphysician published in 1924. Paulsen's error is avoided at least in respect of the pre-critical writings. But it is still suspicious that of the later writings the incomplete essay for a Berlin Academy prize competition, *What is the real progress that metaphysics has made in Germany since the time of Leibniz and Wolff?* is given great importance. This essay was not completed by Kant, and in particular was not submitted and not published. Wundt's interpretation relies essentially on this manuscript, and this seems to us not only a ground for criticism but also to conflict with Wundt's own principles of interpretation. Wundt himself says explicitly 'Every interpretation of Kant has to keep to the works in the first instance, and especially, as is natural, to the principal works' (p. 19*n.*).[1] We should like to stress this principle and in particular to say that anything which Kant himself did not publish may only be drawn on as a secondary aid to interpretation. Our criticism is directed primarily against Wundt's underlying conception of metaphysics. The problem is perhaps most clearly expressed in his own words.

> This intuitive understanding, the creative understanding of God, which creates things by thinking them, perceives all reality in its necessary rational connection, which is in turn determined by the unity of the idea. To reflect on this connection is the highest problem of human thought, the true task of metaphysics. So far as it is possible for human understanding, metaphysics tries to consider the world in connection with the idea of, or from the point of view of, the highest purpose for which God has created it. To reflect on the revelation of God in the world, is, here as always, the proper occupation of metaphysics (p. 427).

And perhaps in an even sharper formulation, 'The idea of God and its working out in knowledge of the world is the final goal of the Kantian philosophy, as it is of metaphysics in general' (p. 389). In this conception of Kantian metaphysics what is right and what is wrong are mixed in the most dangerous way. A certain reservation is in fact made. Metaphysics is to know the world as God's creation, but only in so far as this is possible to a human understanding. To speak quite plainly, it is Kant's doctrine that the human understanding cannot do this. Human understanding has knowledge of appearances and not of God's works. To reflect on the revelation of God in the world is the proper occupation neither of metaphysics nor of Kant's philosophy. Wundt elevates the idea of God to the final goal of metaphysics and of Kantian philosophy and is thereby taking up a particular side in a long-standing discussion about the problems of metaphysics.

It may be admitted that the Aristotelian metaphysics starts in Book XII from an attitude which corresponds in essentials to Wundt's conception of metaphysics and which in its later formulations can be called natural theology. Metaphysics is here the doctrine of the highest being and thus the doctrine of God. But the middle books of the *Metaphysics* show clearly that metaphysics can also be regarded as something different, namely the doctrine of being as being (*Met.* III.1.1003a21). It can be shown that the two attitudes are connected in Greek consciousness and for Aristotle. What anything is can be seen only from the best of its kind, and so the being of what is can only be known from the highest being. The problem of metaphysics is thus twofold; on the one hand it is the doctrine of the highest being, on the other hand, the doctrine of being as being. But it must be said all the same that the problem of metaphysics, even for Aristotle, is regarded more and more as doctrine of being as being, and in its later development metaphysics is taken unambiguously in the same direction. In Aquinas's conception of the problem of metaphysics, for instance, the emphasis lies on the doctrine of being as being (*com. in met.* 391–2,384,389,593).

We can now see that to consider metaphysics exclusively as doctrine of the highest being, in practice as natural theology, must lead to an impoverishment of metaphysics, and it is just this impoverishment which damages Paulsen's and Wundt's interpretations of Kant on the systematic side.

With the return of philosophy to metaphysics, the doctrine of being as being has been recognised again as the central problem of metaphysics, and modern systematic metaphysical enquiries have thus led to a deeper understanding of the Kantian metaphysics.

Martin Heidegger's interpretation [1] of the *Critique of Pure Reason*, published in 1929, consists of a running commentary which goes as far as the end of the chapter on the schematism and breaks off before the Principles of Pure Understanding; it thus does not cover the Transcendental Dialectic. The consequence is that there is no commentary on the being of freedom, of God, and of things in themselves. Heidegger confines himself to the interpretation of the being of appearances and the being of the knowing subject, but this very limitation shows the far-reaching importance of the purely ontological approach.

Nicolai Hartmann took up his position on the problem of Kantian metaphysics in a comprehensive essay of 1924.[1] Of the three great themes, it is the problem of freedom that Hartmann takes as the central problem and thus the distinction between appearance and thing in itself also occupies a central position for him. Hartmann is able to show how the problem of freedom necessitates Kant's fundamental ontological distinction and determines its scope. At the same time it seems that to concentrate on the problem of freedom in this way shifts the relative importance of the other groups of problems within certain limits; and although one may well ask whether any interpretation without a shifting of emphases and accents is possible at all, such an interpretation remains the ideal.

In this problem of interpreting the Kantian metaphysics and at the same time keeping the Kantian emphasis as far as possible, Heinz Heimsoeth seems to us to have succeeded best. His most important writings are the two articles for the Kant bi-centenary in 1924.[1] At the beginning of the first article Heimsoeth gives a summary statement of his position:

> Kant's critical fight against the dogmatism of traditional metaphysics was in no way directed against metaphysical problems and enquiries in general, but only against that naturalistically restricted type of metaphysics which is committed by its method to categories of nature and is thus essentially inadequate for any real understanding of the great spiritual realities, the type of metaphysics which has been worked out by constructive rationalism in league with the natural sciences. That is why the Critiques are only the propaedeutic for the system that is to be newly erected, which will not only give the new totality of a priori elements but also an ontology founded on them—begun in the section on the Postulates in the *Critique of Practical Reason*.

Kant's final intention, as Heimsoeth has made quite plain, is directed towards an ontology, a doctrine of being, and it may not even be necessary to wait for the section of the *Critique of Practical Reason* on the Postulates, for the ontology to begin. The *Critique of Pure Reason* is also in the proper sense an ontology.

The task for our interpretation is now clear. We have to regard the Kantian metaphysics not only as a doctrine of the being of God; we must learn to see that to determine the being of space, the being of time, the being of nature, the being of appearances, is as much an ontological problem as to determine of the being of God. The transcendental ideality of appearances and the objective reality of intelligible being are both equally the theme of metaphysics, and the aim of our interpretation is to clarify this double problem.

It is encouraging to notice that Kant himself, looking back on his principal work, regarded this double task as the real difficulty. This is indicated in some of the *Reflections*. 'There are two cardinal principles of all metaphysics; the ideality of space and time and the reality of the concept of freedom' (*Refl.* 6344). In the associated *Reflection* 6348 Kant again calls these two themes the hinges (*cardines*) of critical philosophy. In the Preface to the *Critique of Practical Reason*, which we have quoted above, Kant calls this double relation the real riddle of critical philosophy (V,5; Abb. 90). In this spirit we have to seek the connection between Kant's two fundamental ontological doctrines, the connection between the transcendental ideality of appearances and the objective reality of intelligible being.

§ 21

The Aporiae of the Thing in Itself

The concept 'thing in itself' is introduced as the subject of discussion in the chapter, Of the Ground of the Distinction of all Objects into Phenomena and Noumena. In the first edition, Kant still keeps close to the *Inaugural Dissertation*. In the *Dissertation* Kant had said simply, the senses represent things to us as they appear (*sicuti apparent*), the understanding represents things as they are in themselves (*sicuti sunt*) (II,392). In the first edition of the *Critique* Kant says:

> For if the senses represent something to us merely as it appears, this something must also in itself be a thing and an object of a non-sensible intuition, i.e. of the understanding. That is, a kind of knowledge must be possible in which there is no sensibility and which alone has absolutely objective reality. Through it objects are represented to us as they are, whereas in

the empirical employment of our understanding things are only known as they appear (A 249).

Kant is thus still working here in terms of the opposition of objects as they are and objects as they appear to us. In the second edition however Kant speaks more subtly and cautiously.

None the less if we call certain objects, as appearances, sensible beings (phenomena), through distinguishing the mode in which we intuit them from what they are like in themselves, it is already implied in our concept that either these objects considered in their own nature (although we do not so intuit them), or other possible things which are not objects of our senses at all but are thought as objects merely through the understanding, are, as it were, placed by us in opposition to the sensible beings, and called intelligible beings (noumena) (B 306).

Kant regards this distinction between phenomena and noumena in the first instance as only problematic; he says 'as it were, placed in opposition'. Further, he makes a distinction which is also problematic between things in themselves that underlie appearances, and things in themselves that do not appear to our senses at all but can only be thought through the understanding. This would therefore mean a double limitation on our experience. Things in themselves need not be as we experience them, and it also need not be the case that we are able to experience in sensible intuition all the things that there are, even only in their phenomenal character. In the matter of terminology, Kant also uses the opposition sensible beings and intelligible beings for the opposition phenomena and noumena, and also the further opposition appearances and things in themselves. We shall go into the different shades of meaning later.

This opposition between appearances and things in themselves is one of the points attacked by all opponents of Kant, a crux of all interpreters of Kant, a fundamental theme, in a positive or negative sense, of all Kantians, the German Idealists as much as the Neo-Kantians. It has also become a fulcrum for the recent endeavours to bring out the significance of ontological problems for Kant.

Jacobi was the first to take up the problem. In a treatise[1] of 1787 he put forward the following criticism.

The Kantian philosopher departs wholly from the spirit of his system when he says of objects that they make impressions on the senses, thereby excite sensations, and in this way bring about representations; for according to the Kantian doctrine the empirical object, which is never more than appearance, cannot be present outside us and be something other than a representation.

Of the transcendental object we have not the least knowledge, according to this doctrine, and we are never in any way concerned with it when objects come under consideration; the concept of it is at most a problematic concept (*Works*, 1812*f*, II,301).

A few years later, in 1792, G. E. Schulze takes up this criticism again in the *Aenesidemus*. The argument is much clearer and is directed in particular against Kant's calling the thing in itself a cause after having restricted the concept of cause to appearances. Similarly, reality may not be attributed to the thing in itself when the category of reality has also been limited to appearances (edition of 1911, p. 223*f*.).

In the general argument a very effective sentence of Jacobi's exerted a great influence.

I have to admit that this objection has held me up not a little in my study of the Kantian philosophy, so that for several years on end I had to keep on beginning the *Critique of Pure Reason* from the beginning again, because I was continually confused by being unable to find my way into the system without this presupposition, and with the presupposition being unable to stay in it (*Works*, 1812*f*, II,304).

In response to this criticism attempts began at once, either to remove the concept of the thing in itself from the Kantian philosophy altogether, or at least to re-interpret it so as to remove the apparent contradiction with the Transcendental Analytic. C. L. Reinhold [1] was the first to try to re-interpret the concept of the thing in itself in this sense. There followed a series of further attempts, among them those of J. S. Beck and Salomon Maimon. In this, the first stage of the argument about the Kantian philosophy, it is impossible to say that the followers and opponents of Kant are divided uniformly between the two standpoints. In studying the history of the philosophy of the period, one may easily gain the impression that all Kant's followers reject the thing in itself, while his opponents support their arguments on it; but this is not uniformly the case. The closest circle of Kant's pupils retain the thing in itself, for example Kiesewetter[2] and in particular J. Schultz[3], the authorised commentator on Kant.

In German Idealism, that is, in German philosophy up to the death of Hegel, the rejection of the thing in itself first becomes general. This is true primarily of Fichte, who regards the concept of the thing in itself simply as the Achilles' heel of the Kantian philosophy, and it is true also of Hegel, for whom the thing in itself, as something unknowable, is quite inadmissible. The removal of the concept is however always regarded as a clarification and a purification of the Kantian philosophy.

The same can also be said in general of the Neo-Kantianism of the latter part of the nineteenth century. Here again it is only considered possible to bring out the real foundations of the Kantian philosophy if the concept of the thing in itself disappears. In a sense this is only a re-interpretation, since Cohen retains the term 'thing in itself' but interprets the concept solely on the basis of the whole of experience.[1] Natorp then treats the problem consistently and dispenses with the concept of the thing in itself altogether.

Windelband, in what was, it must be said, one of his very early writings, made an intensive investigation into the different phases of the Kantian doctrine of the thing in itself.[1] The investigation is a systematic and a historical one at the same time, and is typical of the attitude of the Neo-Kantians. Windelband distinguishes four systematic possibilities, which he also regards as four temporally successive standpoints. The first standpoint assumes a world of substances or things in themselves which appears to our sensibility as a world of temporal and spatial experience. In our concepts of the understanding we have an adequate representation of this world in its real being. This is the standpoint of the *Inaugural Dissertation.*

In the second phase Kant comes to the negative conclusion that the understanding attains to no real knowledge, and in particular no a priori knowledge, of things in themselves. Kant assumes the existence but also the unknowableness of things in themselves. This would be the standpoint of the first half of the silent decade that was devoted to the completion of the *Critique of Pure Reason*, thus of about 1772–5.

In the third phase not only the knowableness but also the existence of things in themselves is denied. This denial of things in themselves means, according to Windelband, the real destruction of dogmatism and hence also the climax of the critical philosophy. This extreme standpoint, that outside the process of representation there is nothing about which science need trouble itself, is the gift of the Gods that Kant presented to science. This standpoint would belong in time to the second half of the silent decade, thus to about 1775–8.

The fourth and last standpoint is that of the *Critique of Pure Reason* and of all Kant's later works. The opposition appearance and thing in itself is now made equivalent to the opposition sensible and super-sensible, and man is as a rational being thing in himself, as a sensible being appearance.

This account, delivered with élan by the young Windelband, suffers from the fact that the third possibility, which is supposed to represent critical philosophy proper, is only inferred and is not actually expounded in any of the works. Windelband admits explicitly that Kant's true critical philosophy is not fully expounded in any of his writings (p. 261).[1] This is the Achilles' heel of the arguments against the thing in itself, that a 'real' Kant is constructed who is not actually present in the works.

Bruno Bauch may be regarded as the last of the idealist interpreters of Kant. In his small book of 1916 and his larger book of 1921 Bauch forcefully brings out once more the contradictions in Kant's own account of his philosophy. Kant applies the categories of quantity, causality, and reality to things in themselves and thereby contradicts his own doctrine of the limitation of the categories to the field of possible experience (1916, p. 110; 1921, p. 163*n.*).[1] Bauch seeks an interpretation which would avoid these evident contradictions and yet would leave the concept of thing in itself with some significance. To this end he starts from the distinction between 'our understanding' and 'pure understanding'. In terms of this distinction, the thing in itself is independent of our understanding but not independent of pure understanding; on the contrary the thing in itself is grounded in pure understanding (1916, p. 112). Serious objections may be raised against this attitude. Bauch himself admits that it is a development of what Kant said, though a development which is both possible and necessary.

In the Neo-Kantian movement as a whole Alois Riehl[4] was probably the only one to retain carefully the Kantian concept of the thing in itself. The problem then acquires genuine meaning in the enquiries which are directed to the ontological problems of critical philosophy, which we have briefly sketched at the beginning of this chapter. Friedrich Paulsen gives his attention primarily to questions of content and thus to the existence of things in themselves as things in the intelligible sphere, without discussing the ontological questions as such.

Erich Adickes expounded this problem in 1924 in a special study. Adickes attaches particular importance to giving as complete a survey of the material as possible. The wealth of material that he gives is very convincing but it was not his intention to provide a foundation for and discuss the problems of ontology as a whole. He is satisfied to go back for his foundation to an inner experience of Kant's.

In such an (inner) experience are the strong roots of the concept of the thing in itself. . . . And that experience consists for Kant in feeling a breath of the transcendent in the empirical material of objects of experience, in believing that he can, as it were, perceive immediately in the latter the presence and power of the former, in experiencing it 'as if' it were itself the transcendental (p. 14).[1]

Max Wundt achieved a more broadly based standpoint. We have already indicated that in his book [1] the boundary between philosophy and theology is not always observed. But if we can disregard this, the reality of the thing in itself is brought out very clearly, and Wundt makes especially clear that the objective reality of what exists in itself has its original point of approach in morals. Our criticism of Wundt, that he brings theology and philosophy too close together, can also be made against Gerhard Krüger, who pays particular attention to the problem of being in itself in his enquiries in Kantian ethics.[1]

In the works of Heidegger, Hartmann, Heimsoeth, and Heyse the problem of the reality of the thing in itself is at last brought back into the area of philosophical discussion.

Since Heidegger's book [1] stops at the System of all Principles, the problems of the thing in itself do not as such come under discussion, but Heidegger achieves decisive results as regards the ontological characterisation of man, more particularly in bringing out the finiteness of man as grounded in his combination of receptivity and spontaneity; a conclusion which is very significant for the objective reality of the intelligible subject.

Hartmann emphasises once more that any idealistic interpretation of Kant has to remove the concept of the thing in itself.

The Kantians' *exemplum crucis* has always been the thing in itself. Transcendental idealism does not really admit a thing in itself. The concept has no place in this system, and turns it into realism. If this difficulty was still half hidden in Kant himself, it was made plain in Reinhold's articles. From Salomon Maimon onwards the idealists saw this clearly (especially Fichte and Hegel, but also many Neo-Kantians) and drew the only possible consequence, namely to eliminate it entirely and to declare it a non-entity, a meaningless concept (p. 190).[1]

In opposition to this view, Kant's real importance, as Hartmann sees it, lies in his retention of the thing in itself in spite of all the difficulties. 'The thing in itself is the most eloquent witness for the predominance of the aporetic mode of thought in Kant' (p. 190). Hartmann approaches the problem of things in themselves essentially from the standpoint of gnoseology, that is, from the standpoint of the metaphysical (as opposed to the logical and psychological) part of epistemology. There are not only things that we do not know but things that we cannot know, and there are thus bounds, not to the object but to our knowledge.

There is a boundary which is drawn to our knowledge but not to the object of knowledge, and there is thus an unknowable residue of the object which is only intelligible and not sensible, which never enters the sphere of 'appearance' and of the empirically real, and never becomes a 'phenomenon' but is always a mere noumenon (p. 191).

Hartmann thus considers the Kantian distinction between thing in itself and appearance from the point of view that the categories of knowledge are partially identical with the categories of being. 'The boundary of this identity of the conditions is at the same time the boundary of the knowableness of the object' (p. 193).

In this interpretation the attribution of extraneous attitudes to Kant has probably not been entirely avoided. From this point of view appearance and thing in itself are too homogeneous; phenomenal and noumenal being are too much on the same plane and are in a way two parts of one and the same existence. Hartmann himself draws attention to this difficulty. 'It must be remembered that the object of knowledge is divided by the boundary of knowableness into two heterogeneous parts, which makes the one mere appearance, the other thing in itself' (p. 194). The difficulties of Hartmann's proposed interpretation are primarily that a spatial picture is applied to the relation of appearance and thing in itself. The relation between appearance and thing in itself is understood in terms of the relation that two parts of a surface or of a body have to each other. But it must remain questionable, to what extent the relation between appearance and thing in itself may be interpreted in terms of a spatial picture. This problem of simplification by means of a spatial picture is present, it must be admitted, in Kant himself. Kant likes to call appearance and thing in itself the two sides of one and the same object. But if this picture has its difficulties, they are only increased by the simplifying picture that Hartmann uses.

The same objection can be made against Heyse. He uses the more general concept of spheres and regions and this permits him to speak of the phenomenal sphere and its regions (§§ 45–85),[1] but it must not be forgotten that sphere and region are also spatial concepts which would presumably lose their meaning if the spatial basis of these concepts were completely removed.

The most far-reaching enquiries in the problem of metaphysics in general, as in the question of the thing in itself, are those of Heimsoeth. The following considerations will often start from Heimsoeth's enquiries, and a preliminary short account of them is therefore unnecessary.

§ 22

Phenomena and Noumena in the 'Critique of Pure Reason'

The difficulties that arise from the distinction between appearances and things in themselves are certainly not small, and we have discussed the aporiae at length. Objections have been brought against this distinction, both on general systematic grounds and on the particular grounds of the Kantian philosophy itself. It must however be emphasised that Kant himself kept to the distinction without wavering. In the Transcendental Analytic he gives up the standpoint of the *Dissertation*, and the concepts of the understanding or categories apply to appearances and not to things in themselves. Kant thus arrives at the standpoint which Windelband brought out so sharply: the existence of things in themselves combined with their unknowableness.

In the *Critique of Pure Reason* the distinction between thing in itself and appearance is expounded in this sense in the Transcendental Aesthetic, the Transcendental Analytic, and the Transcendental Dialectic equally. Towards the end of the *Critique*, in the Transcendental Doctrine of Method, the positive significance of the thing in itself, which is the subject of the *Critique of Practical Reason*, is worked out in outline. There is, lastly, the summary account of the subject at the end of the Transcendental Analytic, under the title, Of the Ground of the Distinction of all Objects in general into Phenomena and Noumena.

It is true that there is a way of developing the transcendental ideality of space and time without going back to the concept of the thing in itself for anything essential. Rudolf Lehmann drew attention in 1878 to the fact that the transcendental ideality of space and time is maintained by precisely those philosophers who incline to an idealistic interpretation of Kant and deny the concept of the thing in itself.[1] We ourselves made no essential use of the concept of the thing in itself in our account of the Kantian doctrine of space and time. But this only shows at most that the Kantian arguments can be used in support of what may be a purely idealistic interpretation, and it does not show that the distinction was not essential for Kant himself even in the Transcendental Aesthetic. Kant starts immediately from beings in themselves in his opening question, 'What, then, are space and time? Are they real beings? Or are they only determinations or relations of things, but yet such as would belong to them in themselves even if they were not intuited, or . . .' (A 23, B 37). That is to say, Kant begins immediately from relations which are attributable to things even if they are not intuited and hence belong to them in themselves. Space and time are not relations attributed to things in themselves, and correspondingly the final decision runs, 'Space does not represent any property of things in themselves nor does it represent them in their relation to one another . . .' (A 26, B 42).

In accordance with this, Kant can emphasise in his argument with Newton and Leibniz that space and time 'are not things in themselves' (A 39, B 56). At the end of the section on space, Kant points out explicitly that all spatial objects are mere representations of our sensibility, and that 'the true correlate of sensibility, the thing in itself, is not known and cannot be known through these representations . . .' (A 30, B 45). At the end of the section on time Kant again emphasises that all temporal representations concern 'the object merely as appearance' (A 36, B 53). At the end of the Transcendental Aesthetic in the first edition he again points out that much can be said, even a priori, about appearances but 'nothing whatsoever about the thing in itself which may underlie these appearances' (A 49, B 66). The remarks added in the second edition, the last of which refer to natural theology and *intuitus originarius*, do not weaken but rather intensify this thesis.

The Transcendental Analytic is based throughout on the distinction between thing in itself and appearance. We saw that in the chapter on the Schematism, the objective validity of the categories is grounded on the objective validity of pure intuition, in particular of time. The reference of space and time to the distinction between thing in itself and appearance must therefore be carried over directly to the schematised categories and thereby to the principles. Through the necessary schematisation of the categories all employment of the categories is restricted to temporal conditions, although the general opinion—and that of Kant himself in the *Dissertation*—is that the pure concepts of the understanding apply to things in themselves. Kant raises this question explicitly at the end of the chapter on the Schematism. 'The conclusion is drawn that the categories, in their pure significance, apart from all conditions of sensibility, ought to apply to things in general, as they are, and not, like the schemata, only represent them as they appear' (A 147, B 186). The pure categories without sensible schemata have only a logical significance; taken this way they can no longer be said to possess objective meaning. 'Such a representation I can put to no use' (A 147, B 186), and the categories thus only have a viable meaning if they are referred to appearances.

Finally, in the Transcendental Dialectic the distinction between appearance and thing in itself is made the foundation of the enquiry. The contradictions that occur only arise if the distinction between appearance and thing in itself is disregarded and they disappear as soon as it is given proper consideration.

The positive significance of the thing in itself begins to appear in the Transcendental Doctrine of Method, and we devote the next section to this aspect of the problem.

The problem as a whole is summarised by Kant at the end of the Transcendental Analytic, in Chapter III of the second book, under the title, Of the Ground of the Distinction of all Objects in general into Phenomena and Noumena. We have seen that this distinction underlies the whole of the *Critique of Pure Reason*, and it is thus understandable that when the distinction comes to be treated for its own sake, what is said has the form of a summary:

> Although we have already given a sufficient answer to these questions in the course of the Analytic, a summary statement of their solutions may nevertheless help to strengthen our conviction, by bringing all their moments to bear on one point (A 236, B 295).

The doctrine of phenomena and noumena is presented as such a bringing to bear on one point. Kant summarises the fundamental idea of the chapter on the Schematism in the form of the proposition that the categories may only be used when they are referred to sensibility (A 236, B 295). In the first edition Kant then introduces the distinction with the following definitions.

> Appearances, so far as they are thought as objects according to the unity of the categories, are called phenomena. But if I postulate things which

are merely objects of the understanding and which nevertheless can be given as objects of an intuition although not of sensible intuition (being *coram intuitu intellectuali*), such things would be called noumena (*intelligibilia*) (A 248).

Such things must indeed be thought necessarily, 'for if the senses merely represent something to us as it appears, this something must also in itself be a thing and an object of a non-sensible intuition, i.e. of the understanding' (A 249). But the question is whether, 'according to the present constitution of our understanding' (A 250), we can know anything at all about such a thing in itself. The concept which arises in this way is only a 'concept of a noumenon which is not in any way positive and not in any way determinate knowledge of a thing' (A 252), and thus yields no positive knowledge, which alone is our concern. This first edition definition is problematical because the thing in itself is first defined only as the correlate of appearance, and then the same thing is immediately also defined as the object of a non-sensible intuition. The correctness of this inference must always be in doubt. The distinction is therefore introduced more cautiously and subtly in the second edition.

None the less, if we call certain objects, as appearances, sensible beings (phenomena), through distinguishing the mode in which we intuit them from what they are like in themselves, it is already implied in our concept that either these objects considered in their own nature (although we do not so intuit them), or other possible things which are not objects of our senses at all but are thought as objects merely through the understanding, are, as it were, placed by us in opposition to the sensible beings, and called intelligible beings (noumena) (B 306).

Here Kant distinguishes carefully between two possibilities: things in themselves in so far as they underlie or may underlie what is intuited by us, and things in themselves which may perhaps never become objects of our sensible intuitions as appearances. There is, correspondingly, a consistent distinction in the second edition between a noumenon in the negative sense and a noumenon in the positive sense. A noumenon in the negative sense is a thing 'in so far as it is not the object of a sensible intuition'; a noumenon in the positive sense is an 'object of a non-sensible intuition' (B 307). But no non-sensible intuition is given to us, and in both editions Kant can therefore say, 'but what is called by us noumenon, must as such be taken only in a negative sense'. In spite of this, the concept of the noumenon is still 'not only admissible, but as a concept setting limits to sensibility is unavoidable' (A 256, B 311).

The concept of the thing in itself accordingly remains a problematic concept. It is always thinkable that there might be things in themselves; we cannot say that the concept is unthinkable and we must also therefore keep open the possibility of real existence. We can now see that in every part of the *Critique of Pure Reason* the concept of the thing in itself is regarded by Kant as merely problematic. It is certainly true of the Transcendental Aesthetic and the Transcendental Analytic that the thing in itself is merely problematic in character, and it is also true of the Transcendental Dialectic because in it freedom, immortality, and God are only assured as to their possibility.

There is now the question whether this merely problematic approach to the thing in itself in the *Critique of Pure Reason* is a matter of method or whether it is also a matter of substance. Are we never able to go beyond this problematic approach at all, or is it merely that under the principles of method that are presupposed in the *Critique of Pure Reason* we have to restrict ourselves to the merely problematic approach? As Kant himself goes beyond the merely problematic approach in his later works, this question also passes into the purely historical question whether when he wrote the *Critique of Pure Reason*, Kant already intended to go beyond the merely problematic approach from the point of view of practical reason. We believe that the latter is the case. We believe that when Kant was putting together the *Critique of Pure Reason* he already had in view the ontological thesis of the *Critique of Practical Reason*. At all events, we find nothing incompatible with this attitude in either of the two Critiques. In the *Critique of Practical Reason* Kant still maintains that from the purely theoretical point of view, that is, under the methodological conditions of the *Critique of Pure Reason* only, the thing in itself must be regarded as problematic, and on the other hand in the *Critique of Pure Reason* the possibility of going on to the objective reality of being in itself is always kept open. This is true of the Transcendental Doctrine of Method, in which, in my opinion, the standpoint of the *Critique of Practical Reason* is already given, without any substantial difference. But even in the chapter which we are considering, Kant expresses himself very positively on this problem.

Thus the concept of pure and merely intelligible objects is wholly empty of all principles of their application, because we cannot think of any way in which such objects could be given, and the problematic thought which leaves a place open for them only serves, like an empty space, to limit empirical principles, without itself containing or revealing any other object of knowledge outside the sphere of empirical principles (A 259, B 315).

Here Kant is quite consciously keeping open an empty space. In the later works, Kant can fill this empty space, though with limitations (it never really comes to knowledge properly speaking); but even in the silent decade when the *Critique of Pure Reason* was being written he presumably always had this positive possibility in mind. This is quite certainly true of the concrete problems of freedom, immortality, and the being of God. On these topics too the *Critique of Pure Reason* only reaches a problematic solution, but the problematic solution is not a standpoint of substance but one of method only.

We can therefore summarise the general position as follows: in the *Critique of Pure Reason*, and especially in the section called Of the Ground of the Distinction of all Objects in general into Phenomena and Noumena, the distinction between appearance and thing in itself is treated as merely problematic; but at the same time the objective reality of being in itself, which we have still to demonstrate, is always envisaged as a possibility.

§ 23

Phenomena and Noumena in the Second Critique

The *Critique of Pure Reason* reaches a negative conclusion. Things in themselves remain as an empty space which must be assumed as a possibility but which cannot be reached through knowledge gained from experience. The boundaries of pure reason are the result of the distinction between appearances and things in themselves; pure theoretical reason can know appearances but cannot pass the boundary between appearances and things in themselves.

The boundaries set in the *Critique of Pure Reason* are redefined in the *Critique of Practical Reason*, and the empty space is filled with the objective reality of freedom, of the concept of immortality, and of the concept of God. Kant gives a subtle ontological foundation for this extension of the region accessible to reason, but it is also the product of his deepest personal concern. This concern shows itself when he speaks in the *Critique of Practical Reason* of 'the splendid revelation which is given to us through pure practical reason by means of the moral law, namely the revelation of an intelligible world through the realisation of the otherwise transcendent concept of freedom' (V,94; Abb. 187).

But is not the result laboriously achieved in the *Critique of Pure Reason* nullified by this splendid revelation of the intelligible world? Has not the *Critique of Pure Reason* impressed on us again and again that all knowledge is limited to the region of possible experience, that is, to appearances, and that any going beyond this is a transcending of our potentialities and a stumbling around in empty space?

Now there can be no doubt that this extension, which is made by assuming in a positive sense that there is an intelligible world with objective reality, is not only intended by Kant, but that essential philosophical and existential interests of Kant are at work here. When Kant emphasises again and again that God, freedom and immortality are the essential questions of human reason he is expressing his own fundamental attitude. One can understand why the *Critique of Practical Reason* should have been regarded as the real climax of the Kantian philosophy, and indeed why the historical influence of the Kantian philosophy started from this *Critique*. One can perhaps understand why in ontological interpretations of Kant the objective reality of intelligible being should have become almost the only theme, at least in the case of Paulsen and Wundt. For Kant himself there was no disparity between the two *Critiques*. As we have already indicated, Kant regards the transcendental ideality of appearances and the objective reality of the intelligible world as the two fulcrums of a single unitary philosophy, and everything depends on our being able to regain the point in which these two propositions are joined. Everything depends on our recognising that Kant is necessarily compelled to maintain simultaneously the transcendental ideality of appearances and the objective reality of the intelligible world.

Kant himself discusses the problem explicitly in the *Critique of Practical Reason*. The seventh chapter of Section II has the title, How is it possible to think an extension of pure reason in a practical intention, without at the same time extending its knowledge as speculative? In this chapter Kant at once attacks the heart of the difficulty. In the *Critique of Practical Reason* the objective reality of at least three concepts, God, freedom, and immortality, had been demonstrated. Kant once more asserts explicitly that both results are to stand together, those reached in the *Critique of Pure Reason* and those of the *Critique of Practical Reason*. The *Critique of Pure Reason* had shown that for these three concepts 'no objective reality can be found by theoretical means'; whereas the *Critique of Practical Reason* shows that 'through the practical law . . . there is postulated the possibility of those objects of pure speculative reason and the objective reality which speculative reason could not assure them'. The solution is that this result is indeed an extension, but is 'no extension of speculation, i.e. intended for positive use with a theoretical aim' (V,134; Abb. 232). The three ideas of speculative reason are ideas, but not knowledge. In an appended discussion Kant again says explicitly that this extension does not in any way mean the introduction of theological concepts into physics. 'If we have recourse to God as the author of all things in order to explain natural conditions and their changes this is at least no physical explanation and is always a confession that we have come to the end of our philosophy' (V,138; Abb. 236). Thus in the concluding part of this section of the *Critique of Practical Reason* Kant can summarise explicitly the results of the *Critique of Pure Reason* and emphasise once again the necessity of limiting all knowledge to objects and determining grounds of possible experience. The two together compose a unity, and Kant himself found the clearest expression for this unity.

> It is only through speculative restriction of pure reason and practical extension of it, that reason is brought into that state of equilibrium in which reason in general can be used purposefully, and this example shows better than any other that the way to wisdom, if it is to be assured and not impassable or misleading, must with us human beings inevitably pass through science, but it is not until science is completed that we can be convinced that it leads to this goal (V,141; Abb. 239).

The limitation of theoretical reason and the extension of practical reason are not only able to stand together, they must stand together.

We have now reached the question, in what sense this extension is possible, and we must first discuss some terminological questions. We saw that, from the standpoint of the practical the objective reality of God and the objective reality of finite intelligible beings are assured. In addition, there are the things in themselves which underlie spatio-temporal appearances. In one way, God, finite intelligences, and the substrates of appearances can all be taken together. A whole series of passages can be quoted in which Kant speaks of God and of finite intelligences as things in themselves or at least as beings in themselves, although the term things in themselves is in general confined to the substrates of appearances.[1] An important consideration underlies this usage, in which Kant is consciously differing from

Leibniz. As the substrate of every spatio-temporal phenomenon, in so far as it is living, Leibniz assumes a monad, that is a thinking being. Kant holds this thesis to be indemonstrable, and presumably it is indeed an indemonstrable generalisation. Kant prefers to keep both possibilities open. It may be as Leibniz assumes, that every thing in itself is an intelligible being (perhaps of very low intelligence), but it may also be that there are some underlying substrates which are not intelligible beings. From this point of view one is inclined to restrict the term 'thing in itself' to the substrates of corporeal appearances. On the other hand to adhere strictly to this limited meaning would involve a severe narrowing of linguistic usage, so that one can say that while adhering in principle to the restricted usage Kant departs from it in particular places quite often. We do not intend to deprive ourselves of this rather more free terminology by limiting the term 'thing in itself' to the substrates of inanimate bodies. Where a misunderstanding is possible we shall speak of things in themselves in the narrower sense, meaning the substrates of inanimate bodies.

It is indisputable that Kant makes a whole series of statements about noumenal being, about God, and about things in themselves in the narrower sense. But one has to be very careful and cautious about the details. We can perhaps show this in the case of the principle of contradiction. Adickes decides for the radical thesis, 'On the other hand Kant never doubted that the principle of contradiction has unlimited validity which extends to things in themselves.'[2] In a note Adickes says curtly, 'In the case of the principles of identity, of contradiction, and of the excluded middle, it was of course always taken for granted by Kant that they are also valid for things in themselves.'[3] Wundt holds the same view, but the passages quoted by him in support of so strict a thesis are very thin. Yet Wundt's thesis is at bottom not wrong, if only it is stated with the necessary limitations. In spite of this, we intend as far as possible, when engaged on so subtle a problem, not to go beyond Kant's own pronouncements. The question of the validity of the principle of contradiction for things in themselves might for example be expressed concretely in the form, is the principle of contradiction valid for God? Here we have of course no general agreement, and there are many philosophers and theologians who assume that the omnipotence of God is not limited even by the principle of contradiction. All the same the majority of thinkers incline to the other view, that God's acts and God's being are free from contradiction. Leibniz took up this standpoint in an intensified form. He held not only that God's being and thinking are free from contradiction but further that God does not act without sufficient reason. The same conviction may also be attributed to Kant, but one must bear in mind that Kant avoided giving a decision on this subject. Nevertheless, the Transcendental Dialectic rests on the conviction that the contradictions which appear must be due to our thinking, and if one may invert this proposition, the principle of contradiction would be valid for things in themselves and in particular of God. But here we meet a new difficulty. The principle of contradiction presupposes that the opposites to which the principle is to be applied have characters between which there may or may not be a contradiction. Without such characters the principle of contradiction is quite empty. One could indeed say that in a sense the principle of contradiction is valid for the being and the activity of God, but as we know neither God's properties nor his counsels there is a complete absence of characters between which there could be or not be a contradiction. Whether or not the principle of contradiction is valid of God, of finite intelligences, and of things in themselves, as far as intelligibles are concerned it is entirely empty for us and entirely without meaning. It is still possible that the principle of contradiction might only be a function of discursive thought, and in the absence of an explicit decision by Kant, we decline to decide this question.

We confine ourselves to those characters of things in themselves which Kant asserted explicitly. Of these assertions the concepts of actuality and of objective reality are by far the most important, and for us therefore the primary question is, in what sense can Kant attribute actuality and objective reality to intelligible being?

§ 24

Temporality and Spatiality as Ontological Characters

We have considered Kant's doctrine of the transcendental ideality of space and time in detail in the first part of this book. His thesis can be taken purely negatively: space and time are only forms of appearance. But considered ontologically it also contains a positive assertion: everything that is in space and time is real (in the phenomenal sense). Space and time, for Kant as for the ancients, are ontological characters, and Kant's thesis, as we saw in § 4, when we discussed the treatment of space in Book IV of Aristotle's *Physics*, is a continuation of an old argument. For the sake of brevity we shall use the term 'spatio-temporal' in what follows. Strictly speaking it is only true that all appearances are in time, so that strictly speaking time alone is the decisive character of being. The special position of time, fundamental as it may be,[1] lies outside the scope of our enquiry, and we take our stand on the fact that nearly all temporal appearances also have spatial structures, and that in any case Kant treats space and time throughout as parallel, except in the chapter on the Schematism.

Everything real is in time and may also be in space. The deduction of this fundamental thesis is carried out in the two chapters on the Schematism and on the Principles. In Part I we were primarily concerned with the causal problem and thus with the third group of principles, namely Analogies of Experience, but for our present purposes the principles and the schemata of the fourth group, the modal group, are the most important. We are thus concerned now with the three modal categories, namely possibility, actuality, and necessity, the three modal schemata, and the three corresponding principles, which are called Postulates of Empirical Thought. The problem and the conclusion remain the same. Causality, as a pure concept of the understanding, only acquires meaning for us when it is expressed in the schema of a temporal sequence. Otherwise it is empty for us. This is true in general for all the categories, that they only acquire real meaning when they are limited to the conditions of sensibility, that is, when they are limited to temporal conditions. The modal categories also need to be schematised through temporal determinations. Kant therefore says in the chapter on the Schematism,

> The schema of possibility is the agreement of the synthesis of different representations with the conditions of time in general. . . . The schema is therefore the determination of the representation of a thing at some time or other. The schema of actuality is existence in some determinate time. The schema of necessity is existence of an object at all times (A 144, B 184).

Actuality, to pick out the concept which is most important for us now, means existence at a determinate time, and without such existence at a determinate time the category of actuality cannot claim any objective reality. We can now give this a more general formulation. Everything that is in time is actual, and only that which is in time is actual; time is the criterion of actuality.

What is said here in the chapter on the Schematism is repeated in a corresponding form in the chapter on the System of the Principles. The Postulates of Empirical Thought are:

(1) That which agrees with the formal conditions of experience (according to intuitions and concepts) is possible.
(2) That which is bound up with the material conditions of experience (or sensation) is actual.
(3) That which in its connection with the actual is determined according to universal conditions of experience is (exists as) necessary (A 218, B 265).

The postulates take us a long way into concrete problems. The connection with time is most clearly to be seen in the first postulate. That alone is possible which agrees with the formal conditions of experience, and these conditions of experience are at once more closely specified as the conditions of intuitions and of concepts. Thus agreement with the conditions of intuition in particular is presupposed in the concept of possibility, and since, as we know, a spatial intuition is not always necessary, agreement with the conditions of time as a pure intuition is the presupposition of possibility as such. Only that which corresponds to the conditions of this pure intuition is possible, only that is possible which can be experienced in a temporal intuition. Possibility underlies actuality and necessity, since only that can be actual or necessary which is also possible. This is especially true of actuality, and only that is actual which is sensed at some determinate time or is bound up according to the general conditions of experience with something that is the object of sensation at a determinate time. Both from the chapter on the Schematism and from the chapter on the System of the Principles we reach the general proposition that only that which is in time is actual.

In this proposition Kant is taking up an ancient group of problems, and he was repeatedly occupied with the question as to how it was to be formulated exactly.

We find the first account of the subject in Aristotle's treatment of space in the *Physics*, Book IV. Aristotle shows there that the ancients, i.e. the mythological thinkers and the early physiologers, regarded spatiality as the criterion of actuality. Everything that is in space is actual, and only that which at some time exists somewhere in space is real. A sphinx, Aristotle says, using the stock Greek example, is not real because it does not exist anywhere in space or in time.

Doubts were soon expressed, of course, as to whether spatiality is in fact a thoroughgoing character of being. Plato denies explicitly that spatiality and temporality are marks of genuine actuality. God, the ideas, and the immortal souls are not in space and not in time but yet they are real, indeed they constitute the only true reality. Everything that is in space and time either has no being at all or at least it only has the vague, imperfect being of shadow pictures. From the standpoint of a transcendent doctrine of ideas, being in space and time now becomes the mark of a lower grade of being and perhaps even of not-being in general; real being is non-temporal and non-spatial, and everything that is in space is only a transitory appearance.

The separation of a being regarded as real from all spatial and temporal being acquires a special importance in Christian dogmatics, which has to keep the being of God free from all spatial and temporal determinations. The discussion of the question, in what sense space and time determine reality, runs through philosophy as a whole, and Kant met it all the way. The occurrence of this problem in the correspondence between Leibniz and Clarke was perhaps of particular importance for Kant. In order to maintain the absolute reality of infinite space and infinite time in Newton's sense, Clarke has to bring together somehow the being of space and time and the being of God. Clarke, as we saw, calls space and time properties of God.[2] Leibniz keeps on stressing the difficulties which necessarily arise from this conception and repeatedly insists that God's being must be kept free of all spatial and temporal determinations. Space and time are therefore not absolute realities for Leibniz and not ontological characters of reality. Leibniz, with bold consistency, goes so far as to remove space and time not only from the absolute being of God but also from the absolute being of finite intelligences; monads in their absolute being are not determined spatio-temporally, and thus space and time are not ontological characters even of finite monads.

In emphatic opposition to Leibniz, Crusius declares that space and time are thoroughgoing ontological characters. Every finite existent is as such in space and time.[3]

Everything that is, is in space. Everything that is, is in time. Whether and in what sense these two propositions are correct, is a question that Kant posed explicitly again and again. He saw early that these two propositions do not express the conditions of absolute reality but only the conditions of possible experience. This insight is among those elements in the *Critique of Pure Reason* which are already clearly present in the *Inaugural Dissertation*. Spatiality and temporality, as he says in § 25 and § 27 of the *Dissertation*, are not conditions of absolute reality but only conditions of our intuition. It is a mistake to make them into conditions of absolute reality, and this mistaken ontological assumption is the source of endless difficulties and contradictions.

The problem is treated in the Transcendental Aesthetic with complete clarity.

> The proposition that all things are side by side in space is valid only under the limitation that these things are viewed as objects of our sensible intuition. If, now, I add the condition to the concept, and say that all things, as outer appearances, are side by side in space, the rule is valid universally and without limitation (A 27, B 43).

Kant then says similarly of time:

> We cannot say that all things are in time, because in this concept of things in general we are abstracting from every kind of intuition of things; but this is the condition under which time belongs to the representation of objects. If, however, the condition be added to the concept, and we say that all things as appearances (objects of sensible intuition) are in time, then the principle has its legitimate objective validity and universality a priori (A 35, B 52).

The ancient ontological proposition that everything real is in space and in time has here been reduced to the thesis that all appearances are in space and time. The significance of the problem for Kant also emerges from a series of *Reflections*. Thus Kant says in *Refl.* 4673, for instance, 'Everything is at some place and in some time. I do not say that it is false, only that it does not follow'. In *Refl.* 5342 Kant makes a note of the proposition itself, presumably in order to go into it in a lecture. In *Refl.* 5357 he again alludes to the root of the error. Finally, when Mendelssohn takes up the theme again, Kant at once examines the position taken up by the Berlin philosopher whom he esteemed so highly (V, 100–2; Abb. 195–7).

Kant thus puts the proposition, everything is at some place and in some time, under its necessary conditions, but in so doing gives it its fundamental meaning. The proposition is limited to appearances, but as all reality is simultaneously limited to appearances, the proposition now acquires for the first time ever a genuine meaning. In this limitation to appearances space and time become the genuine characters of being. Reality means the being of appearances, and hence reality in the sense of the Transcendental Analytic is definitively determined by space and time.

Hegel saw this most plainly when he said, 'He stays inside the idea of reality and of being, that reality consists in being sensible existence; Kant never gets outside this idea' (XIX, 578).

Reality means for Kant, as Hegel recognised clearly, sensible existence, and that means temporal existence. Now we have set ourselves to show that Kant's concept of reality goes beyond the reality of appearances, but this makes it all the more important to see as clearly as possible that time is the fundamental ontological character of the objective reality of appearances.

§ 25

The Being of Appearances

As we have seen in Part I, appearances comprehend nature running its course according to law in space and time. This nature includes in particular the objects of inner sense, representations as the object of psychology.

The question of the being of appearances involves the general question of the being of nature and particular questions about the being of space and time. We have discussed these questions in Part I and we now summarise our conclusions again from a purely ontological point of view, in order to be able to bring Kant's doctrine of the being of appearances face to face with his doctrine of the being of things in themselves.

It would be a complete misunderstanding of Kant to interpret his doctrine that nature, space, and time are appearances to mean that nature, space, and time are mere illusion. Kant defends himself explicitly against this conception. 'When I say that the intuition of outer objects and the self-intuition of the mind alike represent the objects and the mind, in space and in time, as they affect our senses, that is as they appear, I do not mean to say that these objects are a mere illusion' (B 69).

Accordingly, Kant gives an explicit refutation of idealism in the second edition. Whether this refutation is completely successful may remain undecided. Kant's intention is however beyond doubt.

In the *Prolegomena* Kant again explicitly defends himself against any interpretation which takes away their reality from appearances and reduces them to mere illusion (IV, 290*f* and 374).

The doctrine of the *Critique of Pure Reason* is that appearances are real. Several synonymous expressions are used for this—the being of appearances, the actuality of appearances, the objective reality of appearances.

That being belongs to appearances may be taken for granted. Our argument is directed essentially against the idealistic interpretation of Kant, according to which being belongs only to appearances and to nothing else.

That appearances are actual has been shown in our discussion of the Postulates of Empirical Thought. This passage goes further and defines actuality as the complex of appearances as a whole. This is one of the essential passages which serves as a ground for the idealistic interpretation that all being and all actuality are to be restricted to appearances.

The third concept, which is the most important one for us, is the concept of objective reality. Strictly speaking the concept of objective reality is only applied by Kant to concepts. Hence one can only speak in a wholly precise sense of the objective reality of the concept of God, and not of the objective reality of God. But in many places Kant uses the ordinary, possibly careless, usage, and we shall therefore do the same.

Objective reality is predicated in the first instance of the categories as fundamental orders of nature.

> . . . thus the understanding . . . is able to think synthetic unity of the manifold of sensible intuition a priori . . . and in this way the categories, as mere forms of thought, obtain objective reality, i.e. application to objects which can be given to us in intuition, but only as appearances (B 150).

Similarly Kant says at the beginning of the doctrine of the Principles, 'The possibility of experience is, then, what gives objective reality to all our a priori knowledge' (A 156, B 195).

In the Transcendental Aesthetic Kant prefers to use the term 'objective validity'. 'What we assert is therefore the empirical reality of time, i.e. objective validity in respect of all objects which may ever be given to our senses' (A 35, B 52).

Here in the Transcendental Aesthetic Kant chooses the term 'empirical reality' and thus maintains at the same time the empirical reality and the transcendental ideality of time. Empirical reality is then contrasted with absolute or transcendental reality, that is to say with the reality which is attributable to things in themselves without regard to possible or actual intuition (A 36, B 53). Space and time have empirical reality but do not have absolute or transcendental reality. Here the term 'empirical reality' corresponds to the term 'objective reality' as it is used in the Transcendental Analytic. On the other hand there is one passage in the second edition where 'objective reality' is used in the sense of absolute reality (B 70). This terminological inconsistency may be regrettable, but the substance of the particular passages is not open to misunderstanding. We use the term 'objective reality', as in the Transcendental Analytic, for the whole region of appearances. We shall thus substitute this more general term for 'empirical reality' as it is used in the Transcendental Aesthetic, and speak of the objective reality both of the categories and of the pure intuitions, and again of the objective reality of nature and of space and time.

In what does the being, the actuality, the objective reality of appearances consist? It is easy to state Kant's answer. Their being is transcendental ideality. What is not so easy, is to grasp the exact ontological sense of this proposition.

There can be no doubt that a great number of passages could be quoted in which Kant says explicitly that appearances are only our representations, or more sharply, that appearances never exist outside us but only in us. He says for instance at the end of the Aesthetic:

> What we have meant to say is . . . that if we as subject, or even only the subjective constitution of the senses in general, be removed, the whole constitution and all the relations of objects in space and time, nay space and time themselves, would vanish; as appearances, they cannot exist in themselves but only in us (A 42, B 59).

And again in the Deduction, 'For as appearances, they constitute an object which is merely in us' (A 129).

These assertions can be given a purely nominalistic interpretation and made to mean that appearances only exist when they are represented; one can conceive an extreme nominalism for which appearances, nature, space, and time only exist when they are actually being represented by a particular subject. But it must remain questionable whether such a nominalistic conception fits Kant's meaning. We shall see later that objections can be raised against this conception from the standpoint of the *Critique of Practical Reason*, but this interpretation must seem doubtful from the purely theoretical standpoint as well.

First, 'in us' and 'outside us' are spatial concepts. But it is not permissible to interpret the being of space itself with simple spatial concepts, so that 'in us' and 'outside us' must at least not be taken spatially.

The problem leads to the old and much discussed question of identifying the subject for which appearances are representations. Is it actual particular subjects, or is it universal human reason, or is it a transcendental subject, perhaps even in a purely logical sense? Then all the spontaneity of reason and all prescribing of laws to nature would be no real spontaneity and no real prescribing but only a metaphorical account of purely logical connections. This question which we met in the first part now recurs under the purely ontological viewpoint with renewed force.

We have to go back to the fundamental problem. Goethe expressed it very clearly by saying that the question is asked, how much our self and how much the outside world contributes to our spiritual existence.[1] Ourself as subject, or in other words reason, *ratio*, has a part in knowledge; in this Plato, Aristotle, Aquinas, and Kant are agreed. None of the great masters succeeded in finding a straight solution to the problem. The difficulties in the problem are such that they will probably never be completely overcome, and it will perhaps always remain in the aporetic stage. Although the unsolved difficulties of the Kantian thesis must be brought out, these difficulties may hardly therefore be regarded as a refutation of the *Critique of Pure Reason*.

If we have to be satisfied with finding that a complete clarification of the nature of the reality of appearances was not achieved by Kant, we still have the firmly established fact that for him being, actuality, and objective reality are attributable to appearances, that is to nature, space, and time. We shall be able to take a further step forward if we now turn to the question of the being of things in themselves. We shall first discuss in detail the being of God, of finite intelligences, and of things in themselves in the narrower sense as the substrates of inanimate bodies, and finally in the last chapter we shall raise once more the question of being as a whole.

§ 26

The Being of God

The Transcendental Analytic shows that the categories may only be applied to objects of possible experience. The modal categories are included in this prohibition and thus the categories of possibility, actuality, and necessity may not be applied to God. God is not an object of the mathematico-scientific knowledge which is brought to bear on the causal flow of events in space and time.

On the other hand the Transcendental Dialectic teaches the possibility of God; and the actuality and necessity of God are demonstrated in the *Critique of Practical Reason*, since the presuppositions of this proof go beyond what is given to pure theoretical reason.

Kant thus applies the concepts of possibility, actuality, and necessity to God, contrary to the results of the Transcendental Analytic. We have here the old aporia of the thing in itself in a concrete form, referred to the being of God; and this suits our immediate purpose, which is to discuss the problem concretely in terms of God and the acting subject. There must be a sense of the categories in which they go beyond the region of appearances, and we hope to clarify this sense by considering the question in a concrete form.

It is apparent at once that other categories, besides the modal categories, and other important fundamental concepts can be applied to God. The concept of cause is one of these. The outcome of the fourth Antinomy (the second dynamic Antinomy) was that both the thesis and the antithesis are true, and thus the thesis is true, 'There belongs to the world, either as its part or as its cause, a being that is absolutely necessary' (A 452, B 480).

The concept of unity can also be applied to God; more precisely the concept of uniqueness. There could never be any doubt of Kant's monotheism, and monotheism means the application of the concept of unity to the being of God.

Finally, among those fundamental concepts, of which it is especially important that they are predicated of God, we find understanding and will.

The question is, therefore, with what right Kant applies the concepts of possibility, actuality, necessity, cause, unity, understanding, and will, to the being of God. We actually have here before us the aporia of the thing in itself in its most concrete and transparent form, and Kant himself draws attention to the problem quite clearly in the Preface to the *Critique of Practical Reason*, calling it the real riddle. 'Not until now do we have the explanation of the riddle of critical philosophy, how the super-sensible use of the categories can be denied objective reality in speculation, and allowed objective reality in respect of the objects of pure practical reason' (V,5; Abb. 90).

It is, however, particularly clear in this concrete problem of the being of God that the difficulties do not arise from Kant's special approach as a transcendental philosopher, but are fundamental difficulties in the subject itself. Kant may have found the best way of expressing this when he said in *Refl.* 6282 that the concept of an unconditionally necessary being is 'the most unavoidable and yet the most inaccessible' concept of speculative human reason. The fact that Kant carefully investigates all the old difficulties of the problem constitutes the special significance of his position. It is because he has done this that he can point out repeatedly that the concepts God, freedom, and immortality are the necessary conclusion of human reason, but can also point out repeatedly that human reason can never adequately reach these its final ends.

Kant's fundamental attitude has to be understood as a reaction against the rationalistic theology of the Enlightenment. The theology and philosophy of the protestant Enlightenment underestimated much too optimistically the ontological difficulties that arise here, and Kant through his subtle enquiries was the first to bring out the fundamentally aporetic nature of the problem.

Leibniz cannot be acquitted of this exaggerated optimism. He greatly overestimated the possibilities of natural theology. The extent of his optimism can be seen in the concrete problem of the relations of the Christian churches. Leibniz apparently believed that the separation of the churches could be brought to an end by diplomatic negotiations. His optimism is also shown in one of the most difficult questions of Christian theology, the question of the theodicy, how can God permit evil? Leibniz not only held this problem to be soluble; he believed that he had solved it fully by his own system, and he believed that his own solution was so easy to understand and so conclusive that he undertook to give an account of it in a popular work.

Leibniz thought he had solved the very difficult question of the relation of the intelligible world to the world of appearances by means of the concept of the pre-established harmony. But it is easily understandable that this excessively simple doctrine was the feature of Leibniz's teaching on which criticism fastened universally.

His optimism is manifest again in his treatment of the ontological problems of the divine being and the proofs of God's existence which he always takes much too lightly. The only barrier which remains standing for Leibniz is the distinction between a finite and an infinite understanding. In problems which are not affected by this distinction a final solution is possible, and in all other problems an unending progress leads nearer and nearer to the goal. Thus in the questions of the fundamental properties of God and in all truths which can be reached by a finite proof, a conclusive and adequate solution can be found.

In Leibniz himself this not unobjectionable optimism is always to some extent compensated by the intuitive genius of a great thinker. But in his followers the callowness and poverty of this optimism appears in a hard clear light.

This is true of Christian Wolff, whose systematic account of Leibniz's philosophy in no way does justice to the difficulties of principle in the problem of the being and of our knowledge of God. It is especially true of the two accounts of Leibniz's philosophy which Kant used for his lectures, Baumgarten's *Metaphysics* and J. A. Eberhard's *Propaedeutic to Natural Theology*.

In his *Metaphysics*, § 862, Baumgarten is satisfied with the distinction between *conceptibilis* and *comprehensibilis*, God is *incomprehensibilis*, but is nevertheless *in se conceptibilis*, and Baumgarten can therefore say 'He can be defined in many ways rightly and indeed, really' (in Kant's works, XVII, 168). The possibility of a purely rational definition of God is here greatly exaggerated; there cannot be many thinkers who hold that a real definition of God is possible. Baumgarten is thus going too far when he rejects in advance as deism and as error the more cautious attitude that the existence of God can be known but not his essence. There must be serious doubts as to whether God can be known in his essence, in the sense defined by Baumgarten.

Eberhard also underestimates the difficulties of principle to the same degree. There is indeed a bitter irony here in that Eberhard is able to go back to one of Kant's early writings in speaking of a possible rational knowledge of God. Eberhard says, '1. From every property of God all other properties can be deduced. . . . 2. Any property of God can therefore be made the basis of a system of natural theology.' He names first Wolff, who starts from the concept of the most perfect being, then Sulzer, who starts from the eternal being, then Baumgarten who starts from the most perfect being again, and finally in the fourth place, Kant, who starts from the necessary being. Kant's early work: *The only possible Ground of Proof for a Demonstration of the Existence of God*, appears here in its rightful place, in association with the rational theology of the Enlightenment (XVIII, 538). But this very conjunction only shows how far Kant had moved from this early work, and how he had grown more and more able to do justice to the difficulties in the problem itself. We shall see in what follows how remote was his position from that of Eberhard in this question.

If the question is asked, what ontological determinations of God did Kant admit in the critical philosophy, the first answer must be a negative one, namely that spatial and temporal characters are not attributable to God. This question of the relation of space and time to the being of God is again one of the old questions of ontology as much as of theology. It enters on a new stage at the beginning of the modern era in that infinity comes to be regarded as an essential and irremovable property of both space and time. It was forgotten that Aristotle had regarded space as finite and Plato had regarded time as finite. There is at this time a certain pathos of the infinite which tends to bring together in a special way the infinity of space and time and the being of God.[1] The statement of the idea of infinity that was most important for Kant was probably that of Newton. We have seen more than once that on the Newtonian assumption of the absolute reality of infinite space and infinite time, it is almost impossible to avoid bringing space and time into relation with God's being. Newton takes up this position himself, and so does Clarke representing Newton in the correspondence with Leibniz, though not always in the same form. But in this correspondence Leibniz points out repeatedly that to refer space and time to the divine being raises insuperable difficulties. We have discussed at length in Part I the significance of these considerations for the problem of the ontological status of space and time. Leibniz teaches the phenomenality of space and time, so that for him they cannot be taken as characterising God.

In the *Critique of Pure Reason* the doctrine of the transcendental ideality of space and time solves the ontotheological problem as well. Space and time do not belong to things in themselves but are forms of appearance, and it follows immediately from this that they cannot belong to the divine being. Kant points out repeatedly that with the doctrine of the transcendental ideality of space and time he has also solved the ontotheological problem. In the additions to the Transcendental Aesthetic in the second edition he writes:

In natural theology . . . we are careful to remove the conditions of space and time from all God's intuition. But with what right can we do this if we have previously made space and time forms of things in themselves . . .

for as conditions of all existence whatever, they would also have to be conditions of the existence of God (B 71).

The first feature of the divine being is therefore stated in the negative proposition that neither spatial nor temporal determinations are attributable to the divine being.

Three fundamental features of God's being, namely understanding, will, and authorship, are taken together in the definition given in the *Critique of Practical Reason*, 'Therefore, the supreme cause of nature, which must be presupposed as a condition of the highest good, is a being which by understanding and will is the cause (and consequently the author) of Nature, i.e. God' (V,125; Abb. 222).

The first is put in the form: God is understanding, God is intelligence. On this, Greek philosophy and Christian theology are agreed, as also is Kant.

The concept of an original divine understanding, which is only introduced problematically in the *Critique of Pure Reason*, receives objective reality in the *Critique of Practical Reason*, and what is merely problematic for pure theoretical reason can now be pronounced objective. God knows the innermost essence of things, he knows things in themselves, he is not bound to sensibility, he creates things by thinking them. The divine understanding is intuitive and not discursive, it is not compelled to think its way step by step through what is to be thought, it intuits everything in one intuition.

But if the divine understanding is an understanding in so remarkable a sense, the questions arise whether the divine understanding and human understanding can both be called understanding in the same sense, and if this is possible, in what sense it is possible. The differences admittedly transcend all bounds. Human understanding is discursive, dependent on sensibility, it does not simply create its objects but has to have something given to it. All these limitations fall away for the divine understanding. The divine understanding is a pure understanding free from all limitations, free from all discursiveness, free from all sensibility, free from all receptivity.

But can we make for ourselves any representation at all of such an understanding? What understanding can mean, we only know from our understanding, and thus a statement about a divine understanding is not a statement in the proper sense but only an analogical assertion. If we understand God to be an intelligence, we do this by analogy with ourselves as intelligent beings. We cannot represent God to ourselves in any other way than as a being which has understanding, but this understanding has to be so completely freed from the limitations of our understanding, that we can only think it by analogy with our understanding.

Kant emphasises repeatedly the unbounded distance between our understanding and the divine understanding. But in order to be able to speak about a divine understanding at all, even only analogically, there must remain some connection between divine and human understanding, otherwise it is impossible even to make an inference *per analogiam*. This appears particularly clearly in a marginal note in Kant's working copy of Eberhard's *Propaedeutic to Natural Theology*. Eberhard writes:

> The finite understanding represents not everything, not simultaneously, and not most clearly. The understanding on the contrary which has no limits represents everything possible, simultaneously, and most clearly. A finite and an infinite understanding are thus heterogeneous magnitudes which have no common unit, and such are all finite and infinite realities (Kant, XVIII, 534).

Kant comments on this, 'If this were so, we should not be able to use the concept of our understanding elevated to completeness to represent to ourselves the divine understanding' (*Refl.* 6260). An inference by analogy from human understanding to divine understanding is only possible, if some connection between the two is assured.

Thus fundamental difficulties arise in connection with this first way of characterising God. But this does not alter the fact that Kant always retains the definition, God is an intelligence.

The second feature of divine being, God is will, only appeared in Christian theology. It may be possible to find some sort of expression of a divine will in the myth of creation in the *Timaeus* of Plato. A will of God can only be expressed in an action of God towards the world, at the least in an act of creation and after that in actions directed towards particular human beings. Thus the concept of a divine will only acquires importance in the Christian consciousness. Augustin established the theological and philosophical sense of this concept. Understanding and will now become the two fundamental features of divine being, and the balance between the two becomes one of the great problems of western theology and philosophy. Kant must certainly be reckoned among those who, in human existence as in the divine existence, did not wish to see the importance of the will neglected.

Here again the question arises whether one can speak of a will of man and a will of God in the same sense. The will of God is without limits and bounds. There is nothing that could determine the will of God from outside, not even laws or maxims. The will of man on the contrary is continuously determined, the empirical will by sensible motives, the pure will by the law. In original experience we only know the will of man, and we can therefore only speak of the will of God by way of analogy. In the problem of the will Kant stresses especially heavily the ontological distance between the human will and the divine will. From this distance there results the inscrutability and the unknowability of the divine will. The divine will is absolutely inconceivable and inscrutable (*Refl.* 6084). It is 'a temerity to attribute the connections in our will to God. Inscrutable are his ways' (*Refl.* 6081).

Through understanding and will—as Kant said in the *Critique of Practical Reason*—God is the author of the world. We have seen how subtly Kant constructs his enquiry. In the Transcendental Dialectic a world created by a divine author (the thesis of the fourth Antinomy) is first established as a purely theoretical or problematic possibility. Practical reason alone can demonstrate the objective reality of the concept. Here again in the idea of creation, Greek philosophy and Christian theology go together. In Aristotle the idea of creation in its intuitive form moves into the background because of his doctrine of the infinite duration of the world, and it is limited to the idea of the prime mover. In Plato the idea of creation is all the stronger. He expounded the doctrine of the creation of the world again in full in one of the latest and most mature dialogues, the *Timaeus*. We have already considered the problem in connection with the question of a beginning of the world. We are now interested in it from the point of view of the fourth Antinomy, to know whether, and in what sense, God can be called the author of the world.

The fourth Antinomy had at all events shown that the concept is free from contradiction and therefore possible. This problematic treatment of the question in the Transcendental Dialectic is however only a question of method. In point of fact Kant never had any doubts about the proposition that God is the author of the world. Thus in the *Critique of Practical Reason* he speaks of the assumption that 'God as universal original being is also the cause of the existence of substance (a proposition which can never be given up without at the same time also giving up the concept of God as the being of all beings and therewith his all-sufficiency, on which everything in theology depends)' (V,100; Abb. 194). The concept of God as the author of the world is also investigated in a positive sense in the *Critique of Judgment*. We can bring out this positive attitude of Kant without constraint because we have taken it upon ourselves to maintain a careful distinction between theology and philosophy. The significance that this question had for Kant emerges particularly well from the investigation into the subject which he carried out in the *Critique of Practical Reason* under the title, Critical Examination of the Analytic of Pure Practical Reason. Kant raises the question whether God can be the cause of substance. The solution is that it is only possible to say this through the critical distinction between thing in itself and appearance. God is not the creator of appearances but the creator of things in themselves.

> Thus, the creation of these beings is a creation of things in themselves . . . Consequently when I say of beings in the world of the senses that they are created, I am in so far regarding them as noumena. As it would be a contradiction to say, God is a creator of appearances, so also is it a contradiction to say that as creator he is the cause of actions in the sensible world as appearances, even though he is the cause of the existence of the acting beings (as noumena) (V,102; Abb. 196).

Kant points our explicitly that this solution is only made possible by 'the separation of time (and space) from the existence of things in themselves, which is carried out in the Critique of Pure speculative Reason' (V,102; Abb. 197).

But in what sense, then, can we call God the author and the creator of the world? If we ask what author and creator mean, we can only give a meaning to these concepts from our own relation to things, that is to say, from our relation to appearances. Our representation of author always means and can only mean that a man has produced a thing. The creation of the world by God is a quite different mode of creation from our production of finite objects. That is why creation is interpreted as a *creatio ex nihilo*, but a *creatio ex nihilo* is something we are fundamentally unable to represent to ourselves, because our production is always a producing out of something.

To call God the creator and author of the world is also therefore only to make use of an analogy. The idea that this characterisation is only analogical is to be found in a whole series of *Reflections*. It is expounded concretely in the lectures on the philosophy of religion.[2]

The idea that it is only an analogical characterisation is also used again and again in the three Critiques. A detailed investigation of this subject, also bringing out the historical connections, would be welcome. Here I may ignore the historical aspect and refer only to the explicit treatment of the subject in the *Critique of Judgment*. Kant expounds in § 90 the proofs of the existence of God which are possible in principle. He enumerates as the second possibility inference by analogy. 'We may also think the causality of the original being in respect of the things of the world as natural ends, after the analogy of an understanding as the ground of the forms of certain products which we call artefacts' (V,465).

In a long note to § 90 Kant explains the concept of analogy as such. Here again he says, 'I can think the causality of the supreme cause of the world on the analogy of an understanding, comparing its purposeful products in the world with the artefacts of man, but I cannot infer these properties in it by analogy' (V,464*n*).

We only know the relation of man to works produced by man. We know that this production rests on the understanding of man. That is why we can think the relation of God as creator to the world as created by analogy with the relation of man to his works. But this way of thinking of God is burdened with the fact that it is an analogy, that the relation of the creator to the world is after all a different one from that of man to the things of art and technology. But we can say all the same that Kant takes account particularly carefully of the difficulties in the concept of analogy.

The three fundamental features of the divine being which we have so far discussed, namely understanding, will, and authorship, are combined in a further fundamental feature, that of spontaneity. God's being is further characterised by spontaneity. Here again, as in all the fundamental determinations, Greek conceptions and Christian consciousness are in agreement. Aristotle brought out that God is pure spontaneity, pure *energeia*, and the same thing has repeatedly been expressed in Christian theology in the form that the divine being is pure actuality and that no potentiality is attributable to it.[3]

For Kant also, pure spontaneity is a fundamental feature of the divine being. It follows immediately from the pure spontaneity of divine thought, of the divine willing, and of the divine creation. The spontaneity of God is wholly pure spontaneity which is obscured by no kind of potentiality, no kind of receptivity.

All these characters come together in the Kantian doctrine of the existence of God. We have seen that from a purely theoretical point of view only the possibility of God can be demonstrated; only the *Critique of Practical Reason* can show in the objective reality of the concept of God the actuality and necessity of the divine being. We are now therefore concerned with the ontological problem, whether and in which sense we can attribute to the divine being the modal predicates of possibility, actuality, necessity, existence, and objective reality. If we confine ourselves to the predicates of actuality and objective reality, it is again the case that these concepts have meaning in the first instance only in the region of appearance. What actuality is and what objective reality is, can be shown in the first instance only by means of appearances. The problem is admittedly complicated by the fact that, as we shall see, in the self-consciousness of the spontaneous subject an existence and a reality are perceived which go beyond the region of appearance. If we disregard this problem for the moment, we shall have to say here again that actuality and objective reality are originally predicated of appearances, and that they can only be transferred to the divine being *per analogiam*.

The only connection between the characteristics of divine being and those of finite being is thus an analogical connection, and it follows immediately from this that God is essentially unknowable. If we ask how any knowledge at all of this different and divine mode of being is possible, we then find a surprisingly close connection with the traditional doctrines of natural theology.

As our first starting point we choose a *Reflection*. In his working copy of Eberhard's *Propaedeutic to Natural Theology* Kant notes:

> The deist attributes all reality *in abstracto* to the *ens summum* but none *in concreto*. What should the theist do, to attribute reality to God *in concreto*? We proceed with the choice of realities *via tam negationis quam eminentiae*, but in the way in which we attribute realities *in concreto* to the highest being *secundum analogiam*, *per analogiam* (*Refl.* 6286).

Realities cannot be thought *in concreto* by mere understanding, but are always infected with conditions of sensibility.

> Thus I shall first *via reductionis* free reality as far as possible from all that belongs to it as phenomenon (*adhaerentibus sensitivis*), for otherwise anthropomorphisms will be produced. But then I shall intensify it *per eminentiam* endlessly as *realitas noumenon* (even if all particular determinations *in concreto* should disappear). (The way of *eminentia* must not be taken before reduction; since human perfections could grow to infinity without becoming differ-

ent in species). But because the cancellation of everything sensitive also cancels the concept *in concreto*, which would turn all theism into mere deism, there remains the way of application by analogy, according to which I admit that I do not know what the divine qualities are like in themselves, but only that they are thought in the same relation to the world as human qualities to their products (*Refl.* 6286).

It would be a special undertaking to show how closely this *Reflection* follows tradition. It could at most be objected that this is only a *Reflection*, and that the possibility must always be kept open that Kant is only discussing what he has written down problematically, without declaring it positively. But in the problem of analogy, this *Reflection* must be understood as the expression of Kant's personal standpoint. It must be remembered that Kant actually applies a whole series of characters to the divine being—we have discussed understanding, will, authorship, spontaneity, actuality, and objective reality—and that the question must necessarily arise, in what sense Kant is permitted to apply such concepts to the divine being. We therefore read the frequent applications of the concept of analogy in the *Critique of Pure Reason* and the *Critique of Practical Reason* in the sense of this *Reflection*, and we find the decisive passage in § 90 of the *Critique of Judgment* which we have discussed above.

Kant retains the three ways worked out by tradition, the *via negationis*, the *via eminentiae*, and the *via analogiae*. All the limitations imposed by sensible conditions, for example those of space and time, must first be removed by way of negation, *via negationis*. The characters of the divine being reached in this way can still only be attributed to it by way of analogy, for we can never know of the properties of infinite reality what they are like in themselves. It is not without interest to note Kant's departures from his text; Eberhard lists three ways, *via negationis*, *via eminentiae*, *via causalitatis*. Kant deliberately replaces the concept of causality by the concept of analogy, so that in this detail Kant is keeping more closely to tradition (XVIII, 554; *cf.* the editor's note).

We have to learn to recognise without constraint that Kant is continuing the tradition of Greek and mediaeval philosophy without a break. Fundamental features of the divine being can be stated, but they are attributable to the divine being only by way of analogy.

It must be kept in mind here that the change in standpoint as regards the question of the origin of concepts also brings about a change in the problem of analogy. Aristotle and Aquinas see the origin of concepts in the spatio-temporal world; the concepts won from it by abstraction can then be applied by analogy to another kind of being, the divine being. Kant on the other hand sees the origin of concepts in the pure spontaneity of the understanding and holds that concepts which are pure in origin can only be applied to spatio-temporal appearances after they have been schematised. This gives a second possible solution to the problem of the attributes of God. It is not the categories as such but only the schematised categories that are limited to appearances, and one can therefore imagine that the pure categories could be applied to God. The divine being could then be thought through these pure categories but not known. There can be no doubt that in many places Kant points out this possibility, and Paulsen and Wundt took this solution to be the real solution. But I believe that the concept of analogy also has a fundamental significance for Kant's solution. There is always the question whether any application at all of the pure categories is possible without an underlying schema, at least in outline. In addition, our problem by no means turns on pure categories such as unity, causality, actuality, but also on more complex concepts, such as understanding, will, objective reality, and in the last three concepts at least, no solution of the problem seems possible without the concept of analogy.

We therefore reach the conclusion that the being of God can be characterised by fundamental determinations, namely unity, causality, possibility, actuality, necessity, understanding, will, spontaneity, objective reality, and that because of the difference between the being of God and the being of men, these terms can only be applied to the divine being in a way which is in some way indirect, either through the empty application of the pure categories, or by the way of analogy.

§ 27

The Being of the Acting Subject

The *Critique of Pure Reason* asks the question, How is freedom possible? Kant's solution rests on the distinction between appearance and thing in itself. As appearance man is subjected to the causal sequence of events in space and time, as an intelligible being he is free. As an intelligible being he can intervene by a free decision of his own in the causal system of natural events and make an original new beginning, and he therefore has causality through freedom. We shall now consider the problem from the ontological point of view and ask what ontological presuppositions are implicit in this solution of the problem of freedom. We are thus taking up Heimsoeth's question, what ontological doctrine underlies the doctrine of freedom?[1]

The bearer of moral decisions we call the acting subject, as at A 539, B 567. This is the most suitable term for our enquiry, although Kant prefers the term 'acting being'. We can therefore ask our question in the form, what is the ontological status of the acting subject?

Man is in the first instance appearance. He has a body which is part of the causal system of nature in space and time. The sequence of representations which is the object of psychology also belongs, in Kant's view, to the causal system of nature. But it is Kant's unshakable conviction that man is not only appearance but also has an intelligible existence which shows itself in the fact of morality. We recall the well-known passage from the Conclusion to the *Critique of Practical Reason* in which Kant expresses this fundamental attitude with the purest splendour and the deepest effect.

> Two things fill the mind with ever new and increasing wonder and awe, the oftener and the more steadily we reflect on them: the starry heavens above and the moral law within. . . . The second begins from my invisible self, my personality, and exhibits me in a world which has true infinity, but which is perceptible only to the understanding, and with which (and thereby with all those visible worlds also) I discern that I am not merely in a contingent connection, as in the world of the senses, but in a universal and necessary connection (V,161; Abb. 260).

In the moral law we have the immediate certainty of a new being, our invisible self, our personality. If we ask about the being of this invisible self, the being of the personality, we are in a sense altering the point of view. Kant's arguments are aimed directly at the fact of freedom, and it was in this direction, at least in respect of its immediate historical influence, that Kant's philosophy unfolded its greatest power. But we are now reflecting on the ontological doctrine which underlies this doctrine of freedom, and we are thereby abandoning the deep and original pathos of what Kant said directly, in favour of sober ontological reflection. But we are convinced that it is only ontological reflection which can reveal the real structure of Kant's doctrine of freedom.

Heinz Heimsoeth was the first to recognise the need for enquiry of this kind. His own investigations start from the *Reflections*, which provide a large amount of material bearing on this question. But he also pursued the problem very minutely in the main works, in particular in the *Critique of Practical Reason* and the *Groundwork of the Metaphysics of Morals*. Max Wundt then pointed out that the same investigation can be continued into the central works,[2] and Heimsoeth himself had been able to show that in their treatment of the problem the main works do not differ at all in substance from the *Reflections*. We are now concerned to carry the enquiry into the *Critique of Pure Reason* itself and to show that the *Critique of Pure Reason* is really Kant's central work, even for the problem of freedom.

Kant's enquiry is founded on two propositions. First, the acting subject is not appearance, but has intelligible being. Secondly, the acting subject is conscious of this, its intelligible being.

There are almost innumerable passages in which Kant asserts the intelligible existence of the acting subject, and consequently a wide variety of terms is used. It is the distinction between appearance and thing in itself which makes it possible to solve the problem of freedom, but in the *Critique of Pure Reason* the distinction is expounded only as problematic in respect of the problem of freedom. The purely theoretical standpoint is that it is possible to rescue both causality and freedom if this distinction is adopted. But the objective reality of the concept which is thus shown to be possible is only assured in the *Critique of Practical Reason*. This way of expounding the problem is however only a matter of method and does not mean any change of attitude between the first and the second Critiques. All the ontological presuppositions of the *Critique of Practical Reason* are already given in the *Critique of Pure Reason*.

The fundamental presupposition, the distinction between appearance and thing in itself, is found in a great variety of turns of phrase and expression. It corresponds to the distinction between phenomenal and noumenal being. Kant speaks of noumenal, intelligible, and supersensible existence and actuality. An extensive list of the terms and where they occur may be found in Erich Adickes's book.[3]

With this fundamental distinction between thing in itself and appearance we now come upon the fundamental aporia of the thing in itself in the problem of the acting subject. How if the acting subject is not an appearance can Kant predicate existence and actuality of it, when the Transcendental Analytic has shown that these predicates only have meaning for appearances?

We have to be clear first that the acting subject not only has noumenal existence but is conscious of this noumenal existence. The acting subject is conscious of its intelligible existence; this proposition seems to contradict the proposition that man only appears to himself in inner sense as appearance. But the contradiction is only apparent and the two propositions can be reconciled completely.

The intelligible existence of the acting subject is first inferred by ontological means. Only under this presupposition is it possible, according to Kant, to save freedom. The antinomy of freedom can only be resolved on the basis of the distinction between appearance and thing in itself. Kant keeps to this position without change throughout the three critiques. In the *Critique of Practical Reason* Kant can therefore base his argument on a report of the results of the Transcendental Dialectic. The postulates of practical reason, he says, had been shown to lead to concepts 'which pure speculative reason could expound as problems but could not solve', and consequently, as was shown in the Antinomies,

> to that of which speculative reason contained nothing but antinomy, the solution of which it could only found on a concept which was indeed problematically thinkable but could not be proved and determined as to its objective reality, namely the cosmological idea of an intelligible world and the consciousness of our existence in it, by means of the postulate of freedom (the reality of which it makes plain by means of the moral law, and with it also the law of an intelligible world, to which speculative reason could only point without being able to define its concept) (V,133; Abb. 230).

The *Critique of Pure Reason* can thus merely show the idea of an intelligible world to be possible, and only practical reason can realise 'the consciousness of our existence' in it. Practical reason therefore goes beyond the problems of ontology to an immediate consciousness of our intelligible existence. That is why the conclusion to the *Critique of Practical Reason* can say, not only of the starry heavens above, but also of the moral law within, 'I see them before me and connect them immediately with the consciousness of my existence' (V,162; Abb. 260). My intelligible being is thus given to me in such a way that I see it, as it were, before me, and there is a direct connection between this seeing and my consciousness of my existence.

I experience myself therefore both as appearance to inner sense and as intelligible being for my immediate self-consciousness.

> . . . since one and the same acting being as an appearance (even to his own inner sense) has a causality in the world of sense which always conforms to

the mechanism of nature, but with respect to the same event, so far as the acting person regards himself at the same time as a noumenon (as pure intelligence in an existence not determinable in time), he can contain a ground which determines that causality according to natural laws but which is itself free from any natural law (V,114; Abb. 210).

The acting person can thus regard himself as noumenon, and the possibility of doing this is grounded in the consciousness of spontaneity. The self-consciousness of the acting person is consciousness of spontaneity. It is in the consciousness of free spontaneity, the consciousness that I can act according to a free decision, that the intelligible existence of my invisible self shows itself.

The consciousness of spontaneity is an immediate fact for Kant; this can be supported by many passages and especially by numerous *Reflections*. But it is a consciousness *sui generis*, and not really knowledge. The account of the problem in the main works is always qualified at once with the appropriate limitations.

If we were capable of a further glance, namely an intellectual intuition of the same subject (which indeed is not granted us, and instead of which we only have the rational concept), then we should perceive that this whole chain of appearances, in regard to all that may in any way concern the moral law, depends on the spontaneity of the subject as a thing in itself, of the determination of which no physical explanation whatever can be given. In default of this intuition, the moral law assures us of this distinction between the relation of our acts as appearances to the sensible being of ourselves as subject, and the relation through which this sensible being itself is referred to the intelligible substrate in us (V,99; Abb. 193).

If we disregard for the moment the question of possible knowledge and only ask about being as such, Kant's standpoint is expressed here with unmistakable clarity. An intelligible substrate underlies the subject as sensible being, and the subject as appearance is absolutely opposed to the subject as thing in itself. The being of the subject as thing in itself is then characterised by spontaneity.

The spontaneity of moral acting is for Kant the manifest foundation of all moral events, and this fundamental spontaneity is independent of whatever ethical points of view may be taken up, including the Kantian standpoint of a purely formal ethics. To whatever degree inclinations, advantages, desires may influence our moral decisions, all these determinations can still be countered with the freedom of moral decision. All these motives can always be opposed with the firm resolve, I will do the right. In this spontaneity of the acting subject the true being of man is revealed, and this true being is called by Kant being in itself or noumenal being. We are immediately certain of this spontaneity of the acting subject in our own experience, an immediate self-consciousness says to us, it is I who am acting.

If we now raise the question, how this intelligible being of the acting subject can be experienced, then admittedly we come up against great difficulties. If true being is spontaneity, then the question whether spontaneity, if it is to be known, does not have to turn from an acting and a knowing into a known, raises great difficulties. We shall return to this question in the following section.

The fact of spontaneity and the further fact of its self-experience is repeatedly emphasised by Kant. 'Transcendental freedom is absolute spontaneity in acting' (*Refl.* 6077). He even says, 'Practical freedom is the consciousness of absolute self-activity' (*Refl.* 6000). The actuality of freedom is not, of course, experienced in inner sense. On the other hand Kant steadily denies in the main works any knowledge of our intelligible being in intellectual intuition. The passage which we have quoted from the *Critique of Practical Reason* is one which shows this very clearly. In the tentative formulations of the *Reflections* Kant occasionally considers intellectual intuition; for instance, 'We cannot infer the actuality of freedom from experience but we have a concept of it through our intellectual inner intuition (not inner sense) of our activity, which can be moved by *motiva intellectualia*' (*Refl.* 4336). From the spontaneity of moral action the spontaneity of all action radiates. 'I do this, does not mean, another brings this about, and even when I say, I suffer this, this means the intuition of a subject which is before itself and suffers' (*Refl.* 4338).

The difficulties in this question are expressed by Kant by saying that although we cannot deny that freedom is possible in the sense of free from contradiction—as was shown in the third Antinomy—and although we can be certain of the objective reality of freedom in self-consciousness, yet we cannot know how freedom is possible in a real sense. Here we come up against the boundary of knowledge. If all knowledge and all understanding is tied to spatial and temporal conditions, then of course it is evident that the objective reality of freedom and of intelligible existence, both of which are not tied to space and time, cannot be understood in this spatio-temporal sense. The results of the Transcendental Analytic could thus stand; we know that freedom is possible but we do not know how it is possible.

On the other hand we can now understand why, for Kant, all

proofs of the existence of God must start from morality and not from nature. Nature is only appearance, and from the transcendental ideality of appearances no inference to the objective reality of God is possible; whereas we are aware of absolute being in our self-consciousness of spontaneity and of intelligible existence, and from this absolute existence of the acting subject it is possible to gain access to the absolute existence of God.

If we summarise once more the concepts which are applied to the acting subject, we find that there is a comparatively large number of them. First, the active subject is always one. Then there is a plurality of acting subjects. This plurality of acting subjects forms a community among themselves, with God as its supreme head. The acting subject possesses causality. Further, it is possible, as is shown in the *Critique of Pure Reason*, and actual, as is shown in the *Critique of Practical Reason*. The categories that are applied to the acting subject are therefore: unity, plurality, substance, causality, community, possibility, and actuality.

In addition there is a number partly of special and partly of general determinations, of the former, understanding and will, of the latter, being in itself, intelligible existence, and objective reality.

§ 28

The Being of the Thinking Subject

We shall try to show that the spontaneity of the knowing subject is identical with the spontaneity of the acting subject, and therefore that the intelligible existence of the acting subject is also attributable to the knowing subject.

It is very plain that this is one of the most difficult problems in the interpretation of Kant. Moreover, we have already established all that we need for our own enquiry to proceed. The intelligible existence of the subject is established on practical grounds, and what we are concerned with in our enquiry is only intelligible existence as such. But we do not wish to avoid the question which must arise concerning the being of the knowing subject.

The question could be solved by starting from the acting subject. The knowing and the acting subject are certainly not two subjects but one and the same subject, for this reason alone, that every moral act involves a knowing, since there can be no freedom without the knowledge and self-consciousness of freedom. From this point of view the ontological determination of the acting subject must carry over to the knowing subject. But one can also ask the question, what is the subject of pure theoretical knowledge? There is always the difficulty that Kant asked and answered the question as to the being of the acting subject in the *Critique of Practical Reason*, but there is no corresponding question and answer about the knowing subject. On this point we therefore have to go beyond what Kant actually said. We are much encouraged to do this by Heimsoeth's enquiry. He also maintains that intelligible existence is experienced in the consciousness of the personality, and supports this first from the knowing subject and then afterwards from the acting subject.[1]

We are thus asking about the subject which underlies what is said in the *Critique of Pure Reason*. What sort of a subject is it, and what sort of a being does it have? Essentially, three answers have been given. It is the empirical subject, it is universal human reason, and it is to be understood purely logically as a transcendental subject or as a transcendental consciousness. The second possibility has never acquired any importance, the first answer was given by Fries and his school, the third answer by Neo-Kantians. If we try to take the subject of pure knowing as the empirical subject, as Fries does, the difficulties are great; whereas solid grounds can be produced for the purely logical conception of the Marburg school. For them every sort of connection is a logical connection, any talk of 'generating' is only a metaphor for dependence on a purely logical ground, and even the conception of origin, which is prominent in the interpretations of the Marburg school, is not to be taken as a temporal origin but only as a way of expressing dependence on a purely logical ground. There can be no doubt that this interpretation succeeds in expressing clearly the objectifying tendency which is an essential feature of the *Critique of Pure Reason*, and no-one can now forget that when Kant speaks of synthesis, connection, and conjunction, the logical interconnections in the system are always included in his meaning. But one is only driven to this exclusively logical interpretation if one has given up the concept of the thing in itself. If, on practical grounds, one has convinced oneself that the concept of the thing in itself and of intelligible existence is a genuine fundamental concept for Kant, then there is the possibility of using the concept of intelligible existence for the understanding of pure theoretical knowing. We shall try to show that the subject of knowing is the individual subject, considered however as pure and therefore as intelligible.

We are thus considering the pure theoretical self-consciousness of the 'I think' from the point of view of the practical self-consciousness of freedom. The fundamental concept of pure self-consciousness as transcendental apperception is developed by Kant in the Transcendental Deduction.

> Now there can be in us no knowledge, no connection and unity between different pieces of knowledge, without that unity of consciousness which precedes all data of intuitions, and by relation to which representation of objects is alone possible. This pure original unchangeable consciousness I shall call transcendental apperception. . . . This transcendental unity of apperception makes, out of all the possible appearances that can ever be together in an experience, a connection of all these representations according to laws (A 107–8).

This representation is the representation of the 'I'. 'But it must not be forgotten that the bare representation "I" in relation to all other representations (the collective unity of which it makes possible) is transcendental consciousness' (A 117*n*).

In the second edition Kant finds the strongest possible expression for this. 'It must be possible for the "I think" to accompany all my representations; for otherwise something would be represented to me which could not be thought at all, and that is equivalent to saying that the representation would be impossible or at least would be nothing for me' (B 131–2).

This 'I think' also means an original spontaneity. All combination is a combining, all unity a unifying. The fact of this spontaneity and the consciousness of this spontaneity underlie the whole development of the argument in the *Critique of Pure Reason*. We may be satisfied with quoting one example. '. . . of all representations combination is the only one which cannot be given through objects, but can only be executed by the subject itself' (B 130).

Heidegger brought out particularly clearly that all unity is unifying, as we showed when we were discussing the problem of unity and unification in Part I. We are now asking not about unity as such but about the subject of unification, the subject of spontaneity. Who is spontaneous, and who unifies?

What we are seeking in spontaneity is the connection between practical and pure theoretical reason. There can be no mistaking that, although when Kant speaks of pure reason as 'originally legislative' (V,31; Abb. 120), he is referring to the fundamental law of pure practical reason, he is consciously taking up the connection with the doctrine and wording of the *Critique of Pure Reason*. It was the aim of the *Critique of Pure Reason* to show that reason is legislative.

This legislative, spontaneous subject cannot be the empirical subject, and Kant said this explicitly in the Transcendental Deduction. But it does not seem to us that we therefore have to follow the Marburg interpretation and at once make the spontaneous subject into a subject which is only to be understood purely logically. There is still the intelligible subject to be considered. Here again we take our lead from a *Reflection*. 'The soul is *substantia noumenon* in transcendental apperception; hence it has no permanence in time, since this can only belong to objects in space' (*Refl.* 6001).

There are in particular three arguments which lead to this standpoint. The first we have already mentioned at the beginning of this section. The subject in its intelligible existence is the carrier of practical spontaneity, and one is therefore inclined to make it also the carrier of theoretical spontaneity.

The second argument might start from the fact that in this way the connection with Descartes (and before him, with Augustin) is established. Descartes' *cogito ergo sum* means that in the consciousness of the 'I think' I experience myself as substance, and it has therefore rightly been transformed into the expression 'I am a thinking thing' though it could perhaps be expressed more accurately in the form 'I am a thinking substance'. If therefore we assume that in the transcendental apperception of the 'I think' the noumenal existence of the knowing subject is experienced, this argument would carry on directly from Descartes.

A third argument is that here the same connection with divine being would appear as in the case of practical reason, over the question of the sense in which understanding and will can be called attributes of God. We saw in the foregoing section that this is possible in the case of will because pure will is experienced as an absolute reality in the self-consciousness of freedom, and that we may therefore make an inference from absolute reality in our own being to absolute reality in the divine being. The same connection must be made the basis in the case of the understanding, if the absolute reality of our intelligible being shows itself in pure understanding. Kant himself set out this argument in a *Reflection*.

It is a necessary hypothesis of the theoretical and practical use of reason as a whole, to make the assumption about our knowledge—and consequently about all purposes and about an intelligible world—that an intelligible world underlies the sensible world, of which the soul as intelligence is the subjective ectype and an original intelligence the cause; i.e. as the noumenon in us is related to appearances, so the supreme intelligence is related to the *mundus intelligibilis* (*Refl.* 5109).

It is no objection to this interpretation, but rather a confirmation of it, to point out that according to Kant's final doctrine, we only know ourselves in inner sense as appearance and not as thing in itself. Heimsoeth has given a full and definitive treatment of this question. Precisely because the knowing subject is spontaneous in its own being, it cannot know itself in this specific being. Pure spontaneity cannot know itself as pure spontaneity. If it knows itself, it is no longer something spontaneous but something known. In our self-observation we are both observer and observed, and this makes it impossible for the pure spontaneity of the observer ever to observe itself as this pure spontaneity.[2]

This necessary limitation of the self-consciousness of the 'I think', which we met in the self-consciousness of freedom, has far-reaching consequences. This self-consciousness of the 'I think' is so to speak only the self-consciousness of a point, it is not knowledge properly speaking. It cannot be real knowledge because in such knowledge the pure spontaneity of knowing would itself have to become the known. This is the point at which Kant can remake the connection with the critical limitation of all knowledge. The Transcendental Analytic limited all knowledge to appearances. We remain in thorough agreement with this if we say that the only point at which we can go beyond the region of appearances is the self-consciousness of freedom and of the 'I think', but also that on this point we neither have nor can have any real knowledge. It is not easy to describe this limitation exactly. On the one hand we cannot know the intelligible being of the acting and knowing subject; on the other hand this intelligible existence is a fact and we experience this fact in immediate self-consciousness. For the most part Kant tries to remove the difficulty by means of distinction between knowing and thinking. We can think this intelligible existence but we cannot know it. We can interpret this to mean that the categories can be applied in pure thought to this supersensible being, but because there is no schematisation of them in pure intuition, these categories in pure thought do not yield any knowledge properly speaking.

We must point out again emphatically that these unbridgeable difficulties of self-experience also apply for Kant to knowledge of God. The existence of God is appreciated through moral facts, and if moral and theoretical self-consciousness is not knowledge, then knowledge in the strict sense of the existence of God is also impossible.

We cannot even grasp the genuine reality of our own understanding and our own will, that is, we cannot have true knowledge of them, and how much less can we have true knowledge of the being of God which is presented to us primarily as understanding and will. Here we must insist again that Kant must not be made into a theologian. The limits which are effective here are irremovable under the fundamental Kantian presuppositions, and they are irremovable for any well-considered thought. Anyone who removes these limits, is not drawing consequences and completing Kant's thought, but is destroying the fundamental assumptions of Kantian thought.

We do not deny that the Kantian solution is extremely subtle and is perhaps balanced on the point of a needle, but who would believe that a solution to this problem could be found which was not alarmingly subtle? Who could name and pronounce the true reality and absolute reality of God, without at least making use of the most extreme subtlety? How many problems are there not, in which the horrible simplifiers, as Nietzsche calls them, make endless trouble, and in what problems could they make more trouble than in these all but inaccessible fundamental questions of absolute reality?

In spite of this, it is not the case that it is impossible here to say anything at all. The knowing subject can be understood by starting from the acting subject. It is one and the same subject which acts spontaneously and thinks spontaneously, and this subject is not exhausted in its appearance in space and time, but as intelligence it also has intelligible being. Finally I would like to emphasise once more that what has been said in this section about the intelligible existence of the knowing subject is not crucial for our fundamental argument. We are only concerned about intelligible existence as such, and the intelligible existence of the acting subject is beyond all doubt.

§ 29

THE BEING OF THINGS IN THEMSELVES

Whatever objections may have been raised against things in themselves by idealistic interpreters of Kant, there can be no doubt that Kant himself speaks of things in themselves. It may not be superfluous to quote his own words in one or two passages. For instance, in the chapter on phenomena and noumena in the first edition of the *Critique of Pure Reason* he says:

> It also of course follows from the concept of an appearance in general, that something which is not in itself appearance must correspond to it, since appearance can be nothing by itself and outside our mode of representation; and hence unless we are to move constantly in a circle, the word appearance must already indicate a relation to something, the immediate representation of which is indeed sensible, but which must be something in itself, even apart from the constitution of our sensibility (upon which the form of our intuition is grounded), i.e. it must be an object independent of sensibility (A 251).

Whether Kant changed his standpoint in the *Opus postumum* is a question which we do not venture to decide. What he would have said in this work if he had been able to complete it, nobody can say. We have made it a principle, only to recognise what Kant said in the works that he published himself and to use everything else only as an aid to interpretation. We have further narrowed this standpoint by taking the *Critique of Pure Reason* as the fundamental and definitive work, although in the problem which we are now discussing, this is unnecessary. Things in themselves are retained by Kant in a constant form with no essential differences in all the works of the critical period. This Kant says in § 32 of the *Prolegomena*:

> In fact, if we regard the objects of the senses, as is proper, as mere appearances, we admit at the same time by so doing that they have a thing in itself as their ground, although we do not know what it is like in itself, but only know its appearance, i.e. the way in which our senses are affected by this unknown something. Thus understanding, precisely by accepting appearances, also admits the existence of things in themselves, and thus far we can say that the representation of such beings which lie at the ground of appearances, and hence are mere intelligible beings, is not merely admissible but is unavoidable (IV,314).

Again, in the *Groundwork of the Metaphysics of Morals*, 'It follows of itself that it must be admitted and assumed that behind appearances themselves there are things in themselves' (IV,451; Abb. 70).

The second edition of the *Critique of Pure Reason* makes no change in principle in the question of things in themselves, and perhaps makes the original standpoint even clearer. In the new Preface Kant says:

But we still make the reservation, which must be duly noted, that though we cannot know these objects as things in themselves, we must at least be able to think them. For otherwise the absurd proposition would follow, that there would be an appearance without anything that appeared (B xxvi).

This position is held consistently in the later works, in the *Critique of Practical Reason*, the *Critique of Judgment*, the polemic against Eberhard, and *Religion within the bounds of Reason alone*, to name only those that are most important for this problem. What is said in these later works about things in themselves always appears as a summary or report of what had been established in the *Critique of Pure Reason*. The following quotation from the *Critique of Practical Reason* is a characteristic passage.

'My enquiries yielded the result that the objects with which we have to do in experience are by no means things in themselves, but only appearances' (V,53; Abb. 143).[1]

The idealistic interpreters of Kant do not really deny that Kant taught the existence of things in themselves, but they regard the doctrine that is actually present as a residue of dogmatism from which transcendental philosophy must be cleansed. According to their interpretation the concept of the thing in itself leads to a contradiction, and only through the removal of the concept can the critical system be made consistent and complete. This interpretation overlooks the fact that the characteristic duality in critical thought is a specific feature of Kant's thought (and not only of Kant's thought). Kant's philosophy rests in fact on two supports, the transcendental ideality of space, time, and nature, and the objective reality of freedom, God, and immortality. This duality in Kantian thought is expressed in the aporiae of the thing in itself and is Kant's characteristic fundamental approach, as is demanded by the problems themselves. That is why Kant could point out at the beginning of the *Critique of Practical Reason* that this is the real difficulty and riddle of critical philosophy (V,5; Abb. 90).

To these things in themselves predicates are attributed which, according to the Transcendental Analytic, may only be predicated of appearances.

Adickes collected an astonishing bulk of material on this point, and we may therefore content ourselves with a summary mention. Things in themselves, we are told again and again, are the causes of appearances, they underlie appearances, they affect our sensibility. Of these concepts, that of affecting is perhaps the most difficult, and fundamental as the concept is for Kant, the difficulties that lie in it have not yet been satisfactorily solved. There can be no doubt that this is what Kant says, and we saw an example of it in the passage quoted from the *Prolegomena*, in the phrase 'the way in which our senses are affected by this unknown something'.

It is said further of things in themselves that they have existence and actuality. This also contradicts the Transcendental Analytic, where existence and actuality, as the eleventh category, may only be applied to appearances.

Adickes is of the opinion that unity and plurality are also predicated of things in themselves.[2] This is only the case if the concept of the thing in itself is taken in the wider sense. Unity is then predicated of God and plurality and community of finite intelligences as a whole. My opinion is that on the contrary Kant is very cautious about using the concepts of unity and plurality in respect of things in themselves in the narrower sense. But this would be a question for special investigation, and does not decisively affect the general ontological problem of things in themselves.

Besides these fundamental concepts which appear in the Table of Categories, there are also other concepts, of which that of corresponding is the most important. 'Doubtless, indeed, there are intelligible entities corresponding to the sensible entities', Kant says in the second edition at B 308. There are things in themselves that correspond to appearances. We discussed the meaning of correspondence from the point of view of appearances in Part I; things in themselves do not need to be as they are presented in appearances. This corresponding now acquires a new significance for fixing the ontological status of things in themselves.

All these Kantian investigations can be summarised in the proposition that things in themselves in the narrower sense also have objective reality. Objective reality is assured, it is true, only for the intelligible substrates in the first instance, first in the fact of freedom for the acting subject, then as a necessary pre-condition of freedom for the divine being. From this point the thesis of objective reality can be extended to cover all things in themselves, including things in themselves in the narrower sense. Here the epistemological factor, that there must be an existence in itself underlying appearances, is combined with the immediate self-consciousness of existence in itself in the sphere of the practical.

The significance of objective reality in respect of things in themselves in the narrower sense resolves itself into four components.

The objective reality of things in themselves means first their independence of our sensibility. Things in themselves are real because they are independent of whether they are actually being known through our sensibility, and thus also of the general form of our sensibility. This independence of being known is the ontological feature of things in themselves which Kant emphasises again and again in the main works and especially in the *Critique of Pure Reason*. Hartmann drew attention with special emphasis to the ontological significance of this independence.[3]

We may regard it as the second component of the objective reality of things in themselves, that they affect our senses. This affecting now has more to do than merely to provide an epistemological explanation of how we reach empirical knowledge. It is also important from the ontological point of view that things in themselves affect us, that substances existing in themselves enter into a causal relation, given in itself, with us as substances existing in ourselves. The epistemic relation, in particular the sensible relation, is at the same time a causal relation. Through this causal relation things in themselves affect knowing intelligences, and thereby make knowledge possible from their side. Heimsoeth made a special enquiry into this causal relation.[4] This affecting presents itself as the exercise of a certain spontaneity, and thus far things in themselves in the narrower sense also participate in the general determination of being in itself, namely spontaneity.

That things in themselves are orderable may be called the third component. This lies admittedly at the extreme limit of what it is possible to say about their ontological status. In spite of the difficulties that arise here, one must insist that the fact that the totality of appearances forms an ordered whole must make possible a certain inference back to things in themselves. Things in themselves, to express the matter concisely, must be orderable. The categories present a totality of possible orders. But it might none the less be the case that this whole apparatus of possible orders was of no use to us, that things in themselves were so tumultuous that they simply could not be ordered by us. Every attempt to express this is burdened with the fundamental aporia of the thing in itself, because anything that one says must be taken from appearances. But under this limitation it is permissible to say that things in themselves are orderable.

Even closer to the boundaries of what is ontologically possible and perhaps beyond them, there is finally a fourth component: things in themselves are works of God. Here I am anxious not to be misunderstood. To say that things in themselves are works of God involves all the difficulties of the concept of things in themselves, and it is completely misunderstood if it is taken as a theological statement. It is a purely ontological, and therefore a purely philosophical statement. There can be no competition between Kant's statement, burdened with all the subtleties of ontology, and the simple statement of *Genesis*, God created heaven and earth.

If one keeps this precaution constantly in mind, Kant's thesis can first be inferred indirectly from the fact that things in themselves are not created by us. They are independent of us, their being does not flow from human spontaneity, neither from thinking nor from willing. If things in themselves are not created by us, their being must go back to God as the author and creator of the world. We take our lead once more from a *Reflection*, and find this thesis expressed in *Refl.* 4135. 'The phenomenon of a thing is a product of our sensibility. God is the author of things in themselves.'[5] In *Refl.* 5981 Kant says, 'But one should not really say that God created appearances, but that he created things which we do not know and arranged a sensibility in us corresponding to them.' Here Kant touches at once on the difficulty which confronts this ontological determination. That which we know, namely appearances, God did not create, and that which God created, namely things in themselves, we do not know.

If we bear in mind constantly the aporetic character of the problem and the consequent impossibility of removing the difficulties, the thesis that things in themselves are works of God may be said to be found in all three *Critiques*.

The doctrine of the third and fourth Antinomies means that the thesis is at least adopted problematically. In the dynamic Antinomies both thesis and antithesis are true, and this means that the thesis that the world has a supreme cause is true. Two things must be borne in mind. First, the thesis is only possible on the basis of the distinction between appearance and thing in itself. Secondly, in accordance with the fundamental approach of the *Critique of Pure Reason*, the thesis is given as merely problematic and merely possible. The thesis, God created the world, is free from contradiction and therefore possible, and cannot be refuted.

The thesis is then demonstrated in the two later *Critiques* in a way which corresponds to the structure of the three *Critiques* as a whole.

If the thesis is only discussed problematically in the *Critique of Pure Reason*, it is pronounced affirmatively in the *Critique of Practical Reason*.

Consequently when I say of beings in the world of senses that they are created, I am in so far regarding them as noumena. As it would be a contradiction to say God is a creator of appearances, so also is it a contradiction to say that as creator he is the cause of actions in the sensible world as appearances, even though he is the cause of the existence of the acting beings (as noumena) (V,102; Abb. 197).

Kant can therefore accept and affirm the thesis that God is also the cause of the existence of substance (V,100; Abb. 194), although only if the distinction between things in themselves and appearances is presupposed. God is not the creator of appearances but of things in themselves.

In the *Critique of Judgment* the development of this argument has progressed so far that the concept of creation can even appear in a paragraph heading. § 84 has the heading, 'Of the Final Purpose of the Existence of a World, that is, of Creation itself.' But even in this new form Kant observes most carefully the limitations which the subject makes necessary. The concept of creation is never constitutive but is always only regulative. If the necessary limitations are carefully observed, the thesis, things in themselves are the works of God, may therefore still be ventured. But it is my belief that Wundt and Krüger have not observed these limitations sufficiently. We admit this thesis as an ontological proposition, but we insist that its ontological and its theological senses must be carefully kept apart. We are concerned here with a difference of standpoint on which discussion is no longer possible. We believe that philosophy is an independent science and that everything is spoiled if the boundaries between philosophy and theology are disregarded. For Wundt critical philosophy culminates in theology, and he can therefore say, 'To reflect on the revelation of God in the world is here as always the proper task of metaphysics'.[6] In a similar sense, and perhaps more strongly, Krüger says that metaphysics is 'an autonomous understanding of the whole of being as the actual creation of God' [7] and a little later, 'Only through the obedient unconditioned asking after moral wisdom, as the moral law demands, is the world knowable in its being in itself as willed by God'.[8] In reply to this thesis of Krüger we can only assert the contrary: being in itself, as willed by God, is unknowable. We cannot know God's will and we cannot know things in themselves, the creation of God. It is always the same difficulty. For the interpretations of Paulsen, Wundt, and Krüger, the *Critique of Pure Reason* is only at bottom a regrettable interlude, from which Kant was fortunately able to find his way back to the truth. We have seen this in Paulsen already. Krüger says in a very similar sense, 'Kant's doctrine was renewed at the point where it was most mortal, in its relation to the fact of science'.[9] The relation to the fact of science is not the most mortal place but the real source of strength in Kant's philosophy. We hope that by bringing out this difference in standpoint, a misunderstanding of what we are now saying may be avoided.

In spite of all this, there is no doubt that the concept of the thing in itself is first introduced by Kant problematically, that it then becomes necessary for epistemological reasons, and that it finally receives its objective reality in the two later *Critiques*. Four components of the reality of things in themselves can be known, even though they are on the boundaries of what is ontologically possible. Things in themselves are independent of men, they affect us spontaneously, they are orderable, and they are works of God.

CHAPTER VI

BEING AS A WHOLE

§ 30

MAN AS APPEARANCE AND AS THING IN HIMSELF

WE have been able to show that there are different kinds of things existing in themselves—there is God, there are finite intelligences, and there are things in themselves in the narrower sense as the substrates of inanimate bodies. In addition to these three modes of being in itself, there is the being that belongs to appearances. We have now reached our concluding question. What is the relation of these different modes of being to each other? In what sense can these different modes be called modes of being?

If we express the question in general ontological terms, it falls into two parts, in accordance with the general division of what exists into appearances and things in themselves. What is the relation of the being of things in themselves to the being of appearances? and what is the relation between the different modes of being in itself; in particular, what is the relation of the being of God to that of finite intelligences and of substrates? If we take the first question first, it has a particular urgency for man. Man is both appearance and thing in himself, he is both phenomenon and noumenon, he has both sensible and intelligible existence. It is therefore a specially urgent question, in what way are these two modes of being related in man?

This Kantian problem brings up all the difficulties with which the body-soul question is burdened. Man has been taken to be a purely material being, or a being containing two substances side by side, or a being whose bodily existence is mere illusion; all possible standpoints have been taken up and no-one has succeeded in removing all the difficulties or even all the essential ones. It can hardly be said that Kant has provided a final solution to the problem, and we will leave unanswered the question as to whether it is possible at all to solve the problem completely. All the essential philosophical solutions are represented in Kant; we find one that could be called Platonic, another that could in a sense be called Aristotelian, and this Aristotelian solution seems in turn to degenerate in some places into a purely nominalistic solution. Different people will judge the matter differently and account it to Kant's credit or discredit, but it seems to us that to offer so broad a solution may not be in all respects blameworthy.

In the Platonic solution the relation of phenomenon and noumenon, and in the concrete case the relation of body and soul, is conceived as a relation of two worlds. This theory of two worlds is perhaps not Plato's final opinion and it ought perhaps to be called Platonistic, but starting from the early dialogues of Plato, and taken up through related Christian ideas, the theory of two worlds has had an immeasurable influence. Leibniz, however earnestly he may have striven to unite Platonism and Aristotelianism, is essentially taking up the standpoint of a two-world theory in his fundamental distinction between a realm of nature and a realm of grace and in the ensuing doctrine of the pre-established harmony. The problematical character of this distinction between two worlds had been thoroughly familiar to Kant since the beginning of his philosophical thinking, but he nevertheless retains the distinction. In the *Dissertation* of 1770 it is given in the title, *Of the Form and Principles of the Sensible and the Intelligible World.* We saw that in this work Kant assumes that there are different forms and principles for the two worlds. The senses know things as they appear, the world of the senses; the understanding knows things as they are, the intelligible world. The difficulties in the proposition, everything that is, is in space and time, are said to arise from the fact that this proposition, which is valid for the world of the senses, is falsely applied to the world of the understanding (§ 27, II,413). The two concepts of *mundus sensibilis* and *mundus intelligibilis*, translated into the German terms 'Sinnenwelt' and 'Verstandeswelt' which are rendered in English by 'world of the senses' and 'world of the understanding', run through all Kant's works, not only the later works but also the *Critique of Pure Reason.* Our present question is, how is man related to these two worlds? He belongs to both of them, and Kant not infrequently expresses this in terms of an explicit two world theory, which we regard as Kant's Platonic conception of the relation of thing in itself and appearance. He writes, for instance, in the *Critique of Practical Reason*, '. . . it is not then to be wondered at if man, who belongs to two worlds, can only regard his own being in respect of its second and highest destiny with veneration and its laws with the highest esteem' (V,87; Abb. 180).

It is clear from this passage that Kant's Platonic conception of the two worlds, like Plato's own conception, has its origin in the problem of the good. To distinguish the world of the senses and the world of the understanding as two worlds is not merely to use an impressive image; the world of the senses is not merely understood purely theoretically as the system of space and time; the world of the senses, considered ethically, is the world of sensuality. Kant's great struggle for action out of pure morality, the struggle which he conducted with such bitterness against all action from sensual motives, is only intelligible if the world of sensuality has a power of its own which can only be overcome through a long struggle and perhaps even never completely overcome. If the world of the senses, the *mundus sensibilis*, is only the way in which we represent something to ourselves, this fight with sensuality is a fight with a shadow.

On the other hand there can be no doubt that Kant did not wish to tear the unity of the world apart into two separate worlds, as he emphasises in the *Critique of Judgment* (Introduction III, Works V,176), and in particular that he did not want to tear man into two parts. Kant insists again and again that it is the same man who belongs to both worlds, that it is the same act even which belongs both to the world of the senses and to the world of the understanding. Kant discusses these difficulties in the last chapter of the Analytic of Practical Reason. He explicitly affirms once more that existence in accordance with the law of natural necessity belongs only to appearances, and that freedom is attributable to the same being as thing in itself, ' . . . but in application, when we try to explain them as combined in one and the same action and thus to explain this combination itself, great difficulties present themselves, which seem to make such a combination impracticable' (V,95; Abb. 189). A little later, Kant explicitly emphasises the difficulties again.

> But another difficulty stands in the way of freedom, so far as it is to be combined with the mechanism of nature in a being that belongs to the world of the senses; a difficulty which, even after all the foregoing is agreed, still threatens freedom with complete destruction (V,100; Abb. 194).

This section taken as a whole shows that Kant saw the difficulties of combining the two worlds very clearly, and that fundamentally he is unable to offer a solution. The problem of *chorismos*, the falling apart of the two worlds, was never overcome in Platonism, and when Aristotle, and indeed Plato himself in the later dialogues, attacked the problem, they had to give up the doctrine of transcendent ideas.

Kant makes his point, however, that a philosophical system which does not recognise the distinction between the world of the senses and the world of the understanding cannot possibly reach a solution of the problem of freedom, because without such a distinction the universal validity of the mechanism of nature makes freedom impossible. How the interconnection of the two worlds is to be understood, in other words how freedom is really possible, lies for Kant beyond our powers of understanding. We can only understand that freedom as such contains no contradiction, and we cannot understand the real possibility of freedom itself. Kant can therefore say that the unsolved and perhaps insoluble difficulty would weigh even more heavily on any other philosophical approach (V,103; Abb. 197). Kant's attitude need not be interpreted as a light-hearted pushing aside of the difficulties; it is a factual indication that the difficulties cannot be finally removed and that the aporetic character of any solution and hence also of Kant's own solution is rooted in the problem itself.

Alongside this solution, which is in a sense Platonic, there is another, which might be called the Aristotelian solution. This would declare that one and the same man presents himself in one respect as an intelligible being and in another respect as a phenomenal being. The reason for calling it an Aristotelian solution is that for Aristotle generic concepts have no separate being, in the sense of a separate supersensible world, and the existence of generic concepts consists for Aristotle only in their being thought. In this case the difference between Kant's sensible and intelligible worlds would only be that one and the same man regarded differently, in different respects, from different points of view, presents himself now as a being of the understanding, now as a being of the senses. Kant says in a passage which we have already quoted, 'Consequently when I say of beings in the world of the senses that they are created, I am in so far regarding them as noumena' (V,102; Abb. 197). And similarly one and the same act, considered in these same different ways, presents itself now as natural necessity, now as absolute spontaneity (V,99; Abb. 194).

But we have to raise serious objections against this Aristotelian conception of the difference between the sensible and the intelligible worlds. How is morality to be understood, if the struggle between sensuality and pure morality is only a struggle between two ways of considering things? I cannot see how this ontological interpretation can do justice to the fact of moral decision.

There is also a further factor. The two worlds are not in every respect of equal status and there is a relation of grounding between them, the world of the understanding underlies the sensible world. This second factor, the grounding of the world of the senses in the world of the understanding, can also only be understood on a Platonic view and not on an Aristotelian view.

If the difference between the two worlds is reduced to a way of considering things, and if this way of considering things is taken in a purely nominalistic sense, the problem as a whole becomes even more unsatisfactory. I should call it a nominalistic attitude if the being of an object is held to consist only in its being represented, and an extreme nominalism would be to say that the being of an object consisted solely in its being actually represented.

There are a whole series of passages in Kant which can only be understood as tending in this direction, in which appearances, and thus space, time, and nature, are said to exist only in our representation. This attitude is however only possible if a purely theoretical point of view is adopted; if one approaches the *Critique of Practical Reason* with this attitude, the world of the senses dissolves into mere representation, and there is no way of seeing how anything can be left of the antithetic of morality.

If one examines the relation between the three possible solutions, it can be said that the expressions which have a nominalistic ring play no very great part on the whole and the attitude that is sustained is the Aristotelian one, which predominates especially in the *Critique of Pure Reason.* The Platonic attitude comes more to the fore in the *Critique of Practical Reason,* while the Aristotelian attitude predominates again in the *Critique of Judgment.*

This means admittedly that in Kant's philosophy this fundamental question only finds an aporetic solution, and there will be many people who will regard this as an objection against Kant's philosophy. For our part, we regard this aporetic solution, which draws on all the essential historical solutions, not as a defect but as a decided advantage.

§ 31

The Analogical Character of the Concept of Being

To complete our account of Kant's ontology, we need to consider the logical structure of the concept of being.

The starting point for the development of the concept is again

Plato. The Platonic distinction between the world of ideas and the world of the senses underlies all the distinctions which we have discussed between intelligible and sensible, noumenon and phenomenon, things in themselves and appearances. Passages can be found in Plato in which being is attributed to ideas alone. Plato would then be taking up the standpoint of Parmenides that only what is, is, and that what is not, is not. Bruno Liebrucks has given an account of this argument, calling it Eleatism, and has shown that the first approach to the doctrine of ideas in the *Phaedo* and in the *Republic* is close to an Eleatic standpoint. He then shows that Plato's further development represents a continuous struggle with this Eleatic component of the extreme doctrine of ideas. The task of the dialectic is to break away from this attitude, and at the height of his dialectical works, in the *Parmenides* and the *Sophist*, Plato succeeds in going beyond the one-sided Eleatism of the extreme doctrine of ideas.[1]

In these dialogues Plato works out the argument that there is an original connection between being and not being, that everything that is, also is not, and that everything that is not, also is.

This doctrine throws a new light on the doctrine of ideas as such. Even if ideas are said to be what truly exists and phenomena what does not truly exist, it is not permissible to deny all being to phenomena as non-existent. In some way, phenomena must be. From this point of view a certain foreshadowing of this dialectical standpoint can be discovered in the doctrine of ideas itself. It may be that phenomena are only shadows, but even shadows are not pure nothingness but are something. It may be that there can be no knowledge that bears on phenomena, but only opinion; but this opinion, even if it is not knowledge, is something.

It is certainly true of the late form of the theory of ideas, as in the *Timaeus* (49*d*), that phenomena must have some sort of being. There is the idea of the elements, for instance the idea of water. There is also the actual water, created by the creator of the world in accordance with this idea of water. Now it may be that this actual water is never pure and clear water, as is demanded by the idea of water, that the actual water is always in some ways impure and mixed with other elements, that it is always somewhat inclined to evaporate and to freeze, but in spite of all this the actual water is still something and not nothing.

If this is generally agreed, that is, if it is impossible to say that ideas have absolute being and that phenomena have absolutely no being, then being also belongs in some way to phenomena, and the question now arises, in what way? Is it the same being which belongs both to ideas and phenomena, or are they related in some other way?

The doctrine of ideas thus runs up against a fundamental ontological problem, and this is the situation in which Aristotle did his philosophical thinking. Aristotle's conflict with the doctrine of ideas is generally regarded as a complete misunderstanding of Plato. Paul Natorp expressed this viewpoint very strongly.[2] But such a misunderstanding would destroy the unity of western thought, and there is no compulsion to take up Natorp's standpoint. Aristotle's philosophy can be regarded as a further development of Plato's fundamental approach, and I hope to establish this fully in a future enquiry. At present only one factor need be considered. Aristotle certainly criticises the doctrine of ideas continuously and sharply. But this criticism is not directed against the assumption that there are ideas as such, but against a particular ontological interpretation of the doctrine of ideas, namely, making it into a transcendent doctrine and hypostatising a second world of ideas somewhere above the heavens. Aristotle's criticism of this doctrine of transcendent ideas, sharp as it is, only bears on the hypostatisation of ideas which, according to him, runs counter to the true meaning of the doctrine of ideas. Aristotle is only carrying forward the criticism of the hypostatisation of ideas which Plato himself had begun in the *Parmenides*. We have seen the doctrine of ideas involves the question, in what sense do ideas and phenomena have being? In taking up the question at this point, Aristotle sees that a hypostatisation of ideas into a world of ideas above the heavens uses too simple a concept of being, and that in such a doctrine, ideas are really only being given the same sort of existence as that of things of the senses.

Aristotle's real objection is that while the real intention of the doctrine of ideas is to reveal a new kind of being, it has not fulfilled this intention, at least in the form of a transcendent doctrine of ideas, and ideas have fundamentally only been given the same sort of being as things of the senses. The point attacked in the criticism is not Plato's assumption that there are ideas, but his failure to reveal a new kind of being for these ideas. The variety within the concept of being shows itself in the existence of ideas, and that is why Aristotle can rightly conclude his criticism of the doctrine of ideas by pointing out the variety within the concept of being (*Met.* I.9.992b18).

In the Platonic distinction between idea and phenomenon a division of being is assumed, and Aristotelian metaphysics, in its fundamental thesis that there is variety within the concept of being, is therefore a genuine continuation of Plato's metaphysics.

Being, according to Aristotle, is neither a univocal nor an equivocal concept, but an analogical concept. A univocal concept is one which is used in all its applications in the same sense. There is an equivocation when a concept is used in different application in completely different senses. The concept of being is neither univocal nor equivocal, but analogical. Aristotle's stock example of an analogical concept is 'healthy'. Originally it is a man that is healthy, but then everything is called healthy that makes or keeps a man healthy, such as a bath or a food or a medicine.

Being, for Aristotle, is not merely articulated, it is so to speak articulated in different dimensions. There is the articulation in the dimension of the ten categories, then in the dimension of transcendentals, then in the dimension of *actus* and *potentia*, and finally in the dimension of truth and falsehood.

The problem of analogy is clearest in the case of the ten categories. Being, in the proper sense, belongs first to substances, but then the other categories, especially quantities and qualities, also have being, and qualities in particular have a real being of their own which is additional to the real being of substance. But in spite of their independent reality, qualities do not have being in the same sense as substances. Qualities have being, but the being of qualities is only analogous to the being of substances.

Aristotle reaches the same result in the case of the distinction between *dynamis* and *energia*. What is possible does not exist in the same sense as what is actual, and here what is actual has true being, by analogy with which what is possible can also be said to have being. There is for Aristotle a wealth of articulations of being, and in them the Platonic distinction merely into idea and phenomenon undergoes an extensive development, is raised to the status of a fundamental ontological problem, and is interpreted by means of the concept of analogy.

Aquinas accepts Aristotle's theses that there is a variety of modes of being and that the structure of the concept of being is analogical. Aristotle's fundamental ontological attitude was probably the principal reason for the movement from Plato to Aristotle, which Aquinas brought about. These two Aristotelian theses prove to be very fruitful for Aquinas's real philosophical concern, the ontological determination of God. Anyone who thinks deeply enough about the problems of fundamental theology will see that the basic doctrines of the theory of ideas can only be applied to God with difficulty. On the other hand Aristotle's doctrine of transcendentals and his doctrine of the analogical structure of the concept of being, can be drawn on for the problems of fundamental theology with very great profit. The distinction between the being of God and that of creatures, which always makes difficulties for a doctrine of ideas, fits effortlessly into the Aristotelian doctrine of the multiple possibilities of being. On the basis of the general variety of modes of being, the difference between divine and human existence, or at least the possibility of a difference, becomes intelligible, and the analogical structure of the concept of being permits the necessary differentiation of divine and finite being. In his exposition of the unity of the attributes of God in fundamental theology, transcendental philosophy and analogical theory combine most happily. The most important attributes are the unity, goodness, and truth of God, and his understanding and will. In transcendental philosophy Aquinas can develop the plurality of the attributes of God without having to introduce real plurality into the being of God; the concept of analogy makes possible a connection between the divine properties and those of finite beings, without removing between them. The concrete argument is most important for our purposes in the case of the understanding and will of God, *intellectus et voluntas Dei*. Neither introduces any real plurality into the divine being. As divine attributes, they are connected only analogically with the corresponding attributes of man, with the human understanding and will. We can only conceive human understanding and will, and we conceive the attributes of God only by means of the analogical connection. If we can speak of an understanding and a will of God, these must issue from our own being and they can only be pronounced, as Aquinas says, *ex creaturis*, 'for we can only name God *ex creaturis*' (*s.th.* I,13,5).

§ 32

Reality and Being

We return to the decisive double assumption of the Kantian philosophy, the transcendental ideality of space and time joined with the objective reality of freedom. Formulated in this way, as Kant formulated it himself, it brings together the most pregnant problems.

Transcendental ideality concerns not only space and time but nature as a whole. Nature is of course, as we have seen, only physically determinable nature, and the world of the living, with its categories, remains a separate problem. On the other hand, objective reality is attributed not only to freedom but also to the concept of God, the concept of immortality, and to things in themselves in the narrower sense as the substrates of inanimate bodies.

When Kant applies the concept of objective reality and therefore the category of actuality to things in themselves in the narrower and in the broader sense, there arises again the ancient fundamental aporia of the thing in itself. According to the Transcendental Analytic, the categories may only be applied to appearances, and in conflict with this, Kant applies a series of categories to things in themselves (in the broader sense). We shall first summarise again the categories applied by Kant to things in themselves, and then in conclusion raise the question, in what sense the categories may be used in this way without running into contradiction with the Transcendental Analytic. This leads to the final question, in what sense Kant's fundamental ontological distinction, that between appearance and thing in itself, is possible at all.

It can be said that nearly all the categories are applied by Kant to things in themselves, in particular the categories of unity, plurality, causality, community, possibility, actuality, and necessity.

Unity is applied to God and to finite intelligences. Kant's monotheism involves applying the category of unity to God; and finite intelligences have unity, since I myself as a thinking and acting being am a unity and I am conscious of this unity so long as I think and act. Whether Kant applies the category of unity to things in themselves as the substrates of inanimate bodies, is in our opinion questionable. Plurality is applied to finite intelligences. There is a plurality of thinking and acting subjects. With things in themselves in the narrower sense the question of plurality remains open, as does that of unity. A number of passages in favour of it can be quoted, but I doubt whether a decided usage can be established. The theoretical possibility that one single thing in itself underlies nature as a whole cannot perhaps be entirely discounted.

Causality through freedom is of course a fundamental concept in Kantian thought. The *Critique of Pure Reason* uses every available means to demonstrate the theoretical possibility of this concept, and the *Critique of Practical Reason* assures the actuality and objective reality of freedom. From this standpoint it is permissible to make the continually repeated assertion that things in themselves are the causes of appearances, refer to things in themselves in the narrower sense, and establish it as a genuine Kantian concept. To this general concept of the causality of things in themselves, the concepts of underlying and affecting are to be added as more special concepts.

Community is first applied to finite subjects which embody in their intelligible community the ancient concept of the kingdom of God. In so far as God is the supreme head of this intelligible kingdom, they are also in community with God. Whether things in themselves in the narrower sense are in community with each other must remain open. Finally, the epistemic connection grounded on the affecting of the subject can be conceived as a community between subject and thing in itself in the narrower sense, in so far as this knowledge is not directed to other intelligences but to inanimate objects.

Possibility is applied to God, freedom, and immortality. It was the aim of the *Critique of Pure Reason* to show in the Transcendental Dialectic that these three concepts are free from contradiction and therefore possible. The *Critique of Practical Reason* adds actuality and objective reality to these concepts, starting from the objective reality of freedom which can be immediately experienced, while the objective reality of God and of the immortal soul can be recognised as necessarily connected with that of freedom. One can say further that Kant also affirms the necessity of the concept of God.

The central concept for us is that of actuality, for which Kant uses a number of synonyms, of which the most important are existence and reality. In the category of actuality the fundamentally aporetic character of the thing in itself again stands out especially sharply. Actuality, as the eleventh category, may only be applied to appearances and actuality is none the less applied to things in themselves. We have followed this out in detail, and we may recall once more the phrase from the *Critique of Pure Reason*, 'But once we have allowed ourselves to assume a self-subsistent reality entirely outside the field of sensibility . . .' (A 566, B 594). Actuality is applied, as we have seen in detail, to God, finite intelligences, and things in themselves in the narrower sense. We find in many places the term 'reality', usually in the strengthened form, 'objective reality'. 'Reality' here is not the category of reality, but the most general ontological term. In this sense of reality, Kant's theory of the transcendental ideality of space and time is put forward in the Transcendental Aesthetic in opposition to the transcendental reality or the absolute reality of space and time.

This term 'objective reality' continues the old ontological term *realitas objectiva*. We regard it as the most general Kantian concept for reality in a purely ontological sense. In Kantian philosophy the general term 'objective reality' is an elucidation of the general concept of being.

In its most general meaning it therefore embraces all the modes of being, and objective reality is said to belong both to things in themselves and to appearances. Accordingly, objective reality is the concept under which are subsumed both the being of appearances, namely the transcendental ideality of space, time and nature, and the objective reality of God, finite intelligences, and things in themselves in the narrower sense. In general the terms 'transcendental reality' and 'absolute reality' are only used by Kant to characterise his opponents' view of appearances which he is attacking, whereas he generally uses the term 'objective reality' to mark the reality of things in themselves.

Our problem reaches its final concentration in this most general ontological term which covers both the transcendental ideality of appearances and the absolute reality of things in themselves. We can now ask, in what sense can objective reality be extended to cover all modes of being, that is, in what sense can objective reality belong both to appearances and to things in themselves, both to phenomenal and to noumenal being? How is this possible at all, and how can it be reconciled with what is said in the Transcendental Analytic?

In this question we meet not only the all but boundless systematic difficulties of this fundamental ontological problem but also in addition the difficulties of method that are peculiar to the Kantian philosophy. Although Kant himself applied the term 'objective reality' both to appearances and to things in themselves, he did not reflect on the possibility of so doing. To express this in purely Kantian terms, we can say that although Kant made the distinction between appearances and things in themselves, he did not reflect on the possibility of the distinction. We shall therefore have to say that Kant did not attain the highest degree of subtlety in discussing the ontological question, subtlety such as is present in the works of Aristotle and his followers. This is in a sense understandable, because the problem of the different modes of being was very much more urgent both for Aristotle and for Aquinas, for Aristotle as the distinction between idea and phenomenon, for Aquinas as the distinction between divine and finite being. But the distinction between appearance and thing in itself still has a fundamental significance in Kant, and if we are to clarify the possibility of such a distinction, we have to go beyond what Kant himself said and the form in which he said it.

If the right to do this is admitted, it can be said that, for Kant too, the notion of objective reality, and therefore being itself, includes the different modes of being only in an analogical way, or at least that the analogical character of the concept of being has an essential importance for Kant.

We have conducted our enquiry in such a way that this general result has been prepared for, as far as possible, by the discussion of concrete questions. The easiest problem to start from is probably that of the sense in which Kant can attribute understanding and will both to man and to God. We have seen that in this problem Kant retains the idea already worked out in ontology that understanding and will can only be attributed both to man and to God in an analogical way.

The same is true of a number of further concepts, for instance that of a creator. If we call God the creator of the world, we are applying a concept which has concrete meaning only for us and only in the world of the senses; we are applying to the relation of God to the world a concept which can only mean for us the production of an object in the world of the senses by a man. But this application can only be made in an analogical way, because the creation of the world by God is a quite different sort of relation, one into which we are not able to claim any insight. We saw that it is in the problems of the understanding and the will that the analogical character of this transference is worked out, in Aquinas as well as in Kant. According to Aquinas, all other concepts which are transferred from created being to divine being can only be transferred in the same analogical way, and we may assume a similar approach as the general foundation of Kant's philosophy. The fundamental concept of unity is a case in which the analogical character of the transference can be brought out particularly sharply. All unity that we know is for Aquinas unity *in creaturis*, for Kant unity in appearances; it is always the unity of a composite and we can therefore say in general terms, for us unity is always the unity of a composite and any other concept of unity is unintelligible to us. But the unity of God is never the unity of a composite; for Aquinas as for Kant it is a unity with no underlying plurality. If we still apply the concept of unity to God, this is no longer the unity which we know, and the unity of God is only analogous to the only unity which we know, that of a composite.

This analogical transference also holds for the most general concept of all, the concept of being as such. Aquinas enunciates this thesis *in extenso*, and we shall also have to assume that it underlies Kant's work. It means that being is only conceivable for us in appearances, *in creaturis*, and when we apply this concept of being to God, we are relying on an analogical connection.

To generalise, we can say that unity, causality, objective reality and being can only be attributed both to appearances and to things in themselves in an analogical way. There is no doubt that this thesis, expressed with this degree of generality, is not to be found in Kant, but it seems to us that the thesis is to be found in his discussion of particular problems, from which the general conclusion can be drawn. It seems to us further that on the basis of the general thesis the fundamental puzzle about the thing in itself becomes understandable, that is, we can now understand why Kant applies the two most general terms of 'objective reality' and 'being' to appearances and to things in themselves.

At the same time we do not wish to overlook that in Kant there is an essentially different situation from that in Aristotle, which derives from their different conceptions of pure concepts. For Aristotle pure concepts are won by abstraction, generally from nature, for Kant they are generated in pure understanding and are only applicable to nature when they have been schematised. There is little difference between the two theories as regards the actual application of these concepts in nature, but for Kant, in contrast with Aristotle, a pure use of the pure concepts grounded in their pure origin is at least thinkable, and this is a possibility which Aristotle does not have. This gives Kant another possible solution, namely that of conceiving the application of the concepts of unity and causality and consequently also of objective reality and being, as a thinking in pure concepts. The application of the concept of unity to God then means that God can be thought as unity but not known. Paulsen considered the general thesis that things in themselves can be thought in pure concepts, but not known, to be Kant's real solution.[1] I do not deny the importance of this conception even in Kant himself, but to appreciate the full breadth of Kant's solution both ideas must be combined, the analogical character of all these applications and the distinction between thinking and knowing.

There is a further problem over which there is a not inconsiderable difference between Kant's situation and Aristotle's. There is a point at which for Kant there can be awareness, in some way, of absolute being itself. We saw that in the self-consciousness of the knowing and acting subject fundamental ideas of Augustin and Descartes are taken up and developed. But this awareness of absolute existence in the self-consciousness of spontaneity is burdened with all the difficulties of being aware of absolute being, and we must not forget that however far-reaching self-consciousness is for Kant, the only possible knowledge is knowledge of appearances.

Only when all these difficulties are kept in mind can the attempt be made to determine the general Kantian sense of being and thus to take our ontological interpretation of Kant's philosophy to its most fundamental problem.

The being of God is said by Kant to be pure spontaneity. The attributes of God, understanding and will, also have pure spontaneity. God's understanding is pure spontaneity because what he thinks he also at the same time creates, and his understanding is not dependent on having what is to be thought given to it. The will of God is pure spontaneity because it acts from pure motivation, and is not dependent on sensible determinations nor even on laws. Generalising this, the being of God, the objective reality of God, means for Kant pure spontaneity.

All the difficulties that we have so far met are involved in this conception. Pure spontaneity, whether as a general attribute of being or in the form of concrete attributes of understanding and will, is inconceivable for us as pure spontaneity. Both the possibilities that we have already mentioned are present in this problem, indeed we expounded them in terms of this problem. We can think the understanding, the will, and the being of God in their pure spontaneity, but not know them, and we can represent the understanding, the will, and the being of God only by way of analogy.

This is also the case in particular with the fundamental characterisation of God as the creator of the world. We saw that for Kant the concept of God as the creator of the world is unassailable. But the creation of the world is a creation through pure spontaneity by the understanding and the will, and purely spontaneous creation is something of which we can have no real knowledge. That is why the Christian tradition calls the creation of the world a *creatio ex nihilo*, which however is something that we cannot represent intuitively.

It is perhaps permissible to say that Kant added special weight to the difficulties which had always been seen. Kant's conception of the being of God as pure spontaneity is a direct continuation of a tradition of Western thought. Pure spontaneity, pure *energia*, is what the divine being is even in Greek philosophy, and it is pure spontaneity, pure *actus*, for the Fathers and for the Scholastics. Kant, while admittedly emphasising more carefully the conceptual difficulties involved, is able to retain the conception of the being of God as pure spontaneity.

The being of man is determined by spontaneity and receptivity together. Man also has spontaneity in so far as any understanding and any will are grounded in spontaneity. But man as absolute being finds himself in the presence of other absolute being, and his thinking and his acting are thus a coming-to-terms necessarily with other already existing being. Man must recognise this other being, and he is therefore necessarily not only spontaneous but also receptive. Divine being is pure spontaneity because there is nothing opposed to divine being, while the being of man is receptive as well as spontaneous because there are always other beings besides the individual man.[2]

This argument can also be expressed by saying that God is the creator of things in themselves and man is the creator of appearances. Man can indeed create, but can only create appearances, while the creation of things in themselves by God shows that God's creation is true creation. The same thing can be put in another way by saying that God is the pure and uncreated creator, while man is both creator and created. If again this is expressed by speaking of the finiteness of man, one has to take care that this finiteness is made to express not only receptivity but also spontaneity.

The spontaneity of human existence is experienced in a peculiar way in the self-consciousness of the 'I think' and the 'I will'. Spontaneity, as we saw, can never be known in the true sense of knowing, it can only be experienced as pure. But our consciousness of spontaneity in pure self-consciousness carries our whole awareness of all pure being and all pure spontaneity. The being of man and in general of all finite intelligences consists in the combination of spontaneity and receptivity, or one could say, man is both creator and created.

The being of things in themselves as the substrates of inanimate bodies, if there are such things at all, is the most remote and the most obscure for us. Only in that things in themselves affect us do we see a slight shimmer of spontaneity. Further, they are created by God but are still visible to us, so that in the being of things in themselves a combination of spontaneity and createdness is visible for us.

If we call both things in themselves and finite intelligences creatures of God, we must insistently reject any attitude which disregards the boundaries between philosophy and theology. Philosophical knowledge always remains burdened with insuperable aporiae. It is when we follow Kantian thought to this extreme point that we have to emphasise that for Kant scientific knowledge by pure reason has an end here. This is true not only for Kant, and we are convinced that the limitations imposed here by Kant signify a factual necessity. In Kantian terminology, we can concentrate the problem in the following aporetic statement. Things in themselves which are works of God we do not know, and appearances which we know are not the works of God. It is very tempting to go a step further here and to say that we have to think a step beyond Kant or that Kant himself thought one step more. This step was consciously taken in German Idealism; Hegel saw quite clearly, more than anyone else, that a positive view of the world as a creation of God means going beyond Kant, and he consciously went beyond Kant in this way. We do not wish to argue with Hegel, but yet it is Kant's really fundamental insight that our thinking, which would like to go further at this very point, here runs up against barriers which it cannot pass. Whoever passes these barriers goes beyond Kantian thought.

The fourth region of being, appearances, is exclusively characterised by receptivity and createdness, and in addition it is only created by man. Appearances are the lowest grade of being, but this has the consequence that appearances are what we are able to know, because we can only know that which we have created.

Thus the being of God is pure spontaneity, the being of finite intelligences and of things in themselves in the narrower sense is spontaneity and receptivity together, and the being of appearances is pure receptivity. Provided that we observe the boundaries of knowledge, we can express the three modes of being through the concepts of creator and created. God is pure creator, finite intelligences and things in themselves are creator and created at the same time, and appearances are only created. There is thus no unitary sense of being, but being is divided into three modes, the connection between which can also be determined in a certain manner.

We do not wish to ignore the fact that in saying this we have reached the boundaries of systematic and historical thinking. Our enquiry must therefore also conclude with Aristotle's words: [3]

Καὶ δὴ καὶ τὸ πάλαι τε καὶ νῦν καὶ ἀεὶ ζητούμενον καὶ ἀεὶ ἀπορούμενον, τί τό ὄν, τοῦτό τίς ἡ οὐσία

NOTES

Preface and Introduction

1. E. Dillmann, *Eine neue Darstellung der Leibnizschen Monadenlehre*, Leipzig 1891, p. 480.
2. G. Martin, *Wilhelm von Ockham*, Berlin 1949.

§ 1

1. *Reflections* 3785, 3902, 5297, 5404, 5650; see also H. Vaihinger, *Kommentar zu Kants Kritik der reinen Vernunft*, Vol. II, Stuttgart 1881, pp. 130*ff*.
2. *Translator's note.* In the original English version of Clarke's side of the Correspondence, Clarke never actually says that space is a property of God, though in the French translation first published (together with the English version) in London in 1717 and repeatedly republished on the continent, he is made to come nearer to saying outright that space is a property of God (4th Reply, § 10). The description of space which he uses most frequently is 'a necessary consequence of God's existence' (3rd Reply, § 3; 4th Reply, § 10; 5th Reply, §§ 36–48), and he also calls space 'the property of an incorporeal substance', suggesting at the same time that besides God there are in empty space 'possibly many other substances which are not matter' (4th Reply, §§ 8, 9).
3. Vaihinger, *Kommentar*, Vol. II, p. 133.
4. G. Martin, *Wilhelm von Ockham*, Berlin 1949, pp. 99*ff*.
5. Suarez, Disp. Met. D. 47 (Opera omnia. ed. André, Paris 1846, XXVI, 781*f*.).

§ 2

1. F. Engel and P. Stäckel, *Die Theorie der Parallellinien von Euklid bis Kant*, Leipzig 1895.
2. L. Nelson, *Bemerkungen über die nichteuklidische Geometrie und den Ursprung der geometrischen Gewissheit*, Abh. d. Friesschen Schule N.F. I, 1906. W. Meinecke, *Die Bedeutung der nichteuklidischen Geometrie*, Kantstudien XI, 1906. P. Natorp, *Die logischen Grundlagen der exakten Wissenschaften*, 2nd ed., Leipzig 1921, p. 309*f*. O. Becker, *Mathematische Existenz*, Jahrbuch f. Philosophie und phänomenologische Forschung VIII, 1927, p. 737. H. J. Paton, *Kant's Metaphysic of Experience*, London 1936, Vol. I, p. 160*f*.
3. G. Martin, *Arithmetik und Kombinatorik bei Kant*, Itzehoe 1939, p. 45*f*.
4. The literature on this topic in Vaihinger's *Kommentar*, I, 253–292.
5. G. Martin, *op. cit.*, pp. 24*ff*.
6. D. Hilbert, *Grundlagen der Geometrie*, 5th ed., Leipzig 1922, pp. 2 and 238.

§ 3

1. On the connection of intuitionism with Kant and the Neo-Kantians, see: A. Heyting, *Die intuitionistische Grundlegung der Mathematik*, Erkenntnis II, 1931, p. 106, and O. Becker, *op. cit.* (§ 2, *n.* 2), p. 738.
2. On the significance of intuition, see the works of the intuitionists *passim*, and, from another angle, there is O. Becker, *op. cit.* (§ 2, *n.* 2), especially pp. 668*ff*. and G. Hamel, *Über die philosophische Bedeutung der Mathematik*, pp. 11*ff*.

3. Lambert, *Theorie der Parallellinien*, ed. Bernoulli, 1786: 'One might almost draw the conclusion from this, that the third hypothesis occurs on the surface of an imaginary sphere.' Quoted in F. Engel and P. Stäckel, *op. cit.*, (§ 2, *n.* 1), p. 203.

4. On the problem of spontaneity see O. Becker, *op. cit.* (§ 2, *n.* 2), *passim*; D. Hilbert, *op. cit.* (§ 2, *n.* 6); G. Hamel, *op. cit.* (§ 3, *n.* 2), p. 7; P. Natorp, *op. cit.* (§ 2, *n.* 2); R. Dedekind, *Was sind und was sollen die Zahlen*, 6th ed., Braunschweig, 1930, p. III; Aristotle, *Physics* IV.14.223a22; H. Weyl, *Die heutige Erkenntnislage in der Mathematik*, Symposion I, 1927, p. 23.—By drawing attention to the spontaneity of mathematics I do not wish to deny its character as given, which is particularly emphasised by Plato. The possibility of a solution must be sought in dialectical thought.

5. G. Martin, *op. cit.* (§ 2, *n.* 3), pp. 45*ff.*

6. On the constructive character of arithmetic, see especially O. Becker, *op. cit.* (§ 2, *n.* 2), pp. 621*ff.*

7. According to present opinion, formal logic also is not demonstrable from the principle of contradiction alone, and is axiomatic and constructive. On the Kantian definition of the term, it must therefore be called a synthetic science. I am in complete agreement here with E. Cassirer, *Kant und die moderne Mathematik*, Kantstudien XII, 1907, p. 37, and the passages quoted there from B. Russell, *The Principles of Mathematics*, 1903, § 434, and *A Critical Examination of the Philosophy of Leibniz*, 1900, §§ 11 and 12. Wittgenstein and more recently many of his followers, especially Carnap and Behmann, have reached a different conclusion through starting from a different definition of the concept 'analytic'.

§ 4

1. P. Natorp, *Platos Ideenlehre*, 2nd ed., Leipzig 1921, pp. 34*ff.*

2. N. Hartmann, *Diesseits von Idealismus und Realismus*, Kantstudien XXIX, 1924, p. 189.

§ 5

1. P. Natorp, *op. cit.* (§ 4, *n.* 1).

2. See also H. Weyl, *op. cit.* (§ 3, *n.* 4), p. 31.

3. N. Hartmann, *Die Grundlegung der Ontologie*, Berlin 1935, pp. 242*ff.*

4. M. Aebi, *Kants Begründung der 'deutschen Philosophie'*, Basle 1947, only became accessible to me after I had finished this book, and I must reserve my discussion of it.

§ 6

1. H. Leisegang was the first to draw attention to this, in an article of 1915 (see § 16, *n.* 4), and the problem was later investigated in a dissertation by H. Knittermeyer, *Der Terminus transzendental in seiner historischen Entwicklung*, Diss. Marburg 1920.

2. See also *Kants Vorlesungen über die philosophische Religionslehre*, ed. Pölitz, 2nd ed., 1830, p. 12.

3. N. Hartmann, *op. cit.* (§ 4, *n.* 2), p. 186.

4. Vaihinger, *Kommentar*, II, p. 275*f.*

5. Vaihinger, *Kommentar*, II, pp. 290*ff.* gives a comprehensive summary of the discussion.

6. Still among the best accounts of the Transcendental Aesthetic are those which were written under Kant's eyes, namely Johann Schulz, *Erläuterung über de*

Herrn Prof. Kant Kritik der reinen Vernunft, 2nd ed., Königsberg 1791, and *Prüfung der Kantischen Kritik der reinen Vernunft*, Königsberg 1791*f*.

§ 7

1. Other literature on this topic is as follows: A. Riehl, *Der philosophische Kritizismus*, Leipzig 1876, Vol. I, p. 335; B. Erdmann, introduction to *Reflexionen Kants zur kritischen Philosophie*, Leipzig 1882; A. Menzel, *Die Stellung der Mathematik in Kants vorkritischer Philosophie*, Kantstudien XVI, 1911, p. 159; P. Bernays, *Über den transzendentalen Idealismus*, Abh. d. Friesschen Schule, N.F. IV, 1918; M. Kowalewski, *Über die Antinomienlehre als Begründung des transzendentalen Idealismus*, Abh. d. Friesschen Schule, N.F. IV, 1918.

2. Aquinas, *comm. in Ar. physicam*, VIII.2 (*Works*, II,367*f*.). On this point I find myself in agreement with Th. Esser's distinguished work, *Die Lehre des Thomas bezüglich der Möglichkeit einer ewigen Weltschöpfung*, Jb. f. Philos. u. spek. Theol., 1891, p. 176.

3. J. Brucker, *Historia critica philosophiae*, Leipzig 1742.

§ 8

1. G. Joos, *Lehrbuch der theoretischen Physik*, 6th ed., Leipzig 1945, p. 490.

2. E. Mach, *Die Mechanik in ihrer Entwicklung hist.-krit. dargestellt*, Leipzig, 1883; G. Hamel, *Theoretische Mechanik*, Berlin 1949, pp. 51*ff*., pp. 507*ff*.

3. C. F. von Weizsäcker, *Das Verhältnis der Quantenmechanik zur Philosophie Kants*, Die Tatwelt XVII, 1941, p. 86 (reprinted in *Zum Weltbild der Physik*, Leipzig 1947).

4. G. Hessenberg, *Grundbegriffe der Mengenlehre*, Abh. d. Friesschen Schule, N.F. I, 1904, p. 633; A. Fränkel, *Einleitung in die Mengenlehre*, Berlin 1919, p. 133.

5. B. Russell and A. N. Whitehead, *Principia Mathematica*, 2nd ed., Cambridge 1935, Vol. I, p. 77.

§ 9

1. Leibniz, *Opera philosophica* ed. Erdmann, Berlin 1840, p. 77a.—*Cum Deus calculat et cogitationem exercet fit mundus.*

§ 12

1. Chr. Wolff, *Ontologia*, 2nd ed., Frankfurt 1730, § 70; A. G. Baumgarten, *Metaphysica*, Halle 1757 (reprinted in the Berlin academy edition of Kant's works, XVII,31); C. A. Crusius, *Weg zur Wahrheit und Zuverlässigkeit der menschlichen Erkenntnis*, Leipzig 1747, § 290.

2. The work of Hamel and that of Hermes are important. G. Hamel, *Die Axiome der Mechanik*, Handbuch d. Physik, ed. H. Geiger and K. Schelle, Vol. V, Berlin 1927, and *Theoretische Mechanik*, Berlin 1949; H. Hermes, *Eine Axiomatisierung der allgemeinen Mechanik*, Forschungen zur Logik, N.F. 3, Leipzig 1938.

3. M. Heidegger, *Kant und das Problem der Metaphysik*, Bonn 1929, §§ 33, 34.

4. K. Reich, *Die Vollständigkeit der kantischen Urteilstafel*, Berlin 1932.

5. H. Cohen, *Kants Theorie der Erfahrung*, 3rd ed., Berlin 1918, p. 314.

6. Cohen, p. 318.

7. This point would also be involved if we were to discuss Kant's relation with Hume, whose importance for Kant I do not in any way underestimate. The real

point at issue between Kant and Hume seems to me to be whether the causal law is to be regarded as a priori or a posteriori. But this raises a serious difficulty which, in my opinion, Kant greatly underestimates when he assumes that every proposition is either a priori or a posteriori. In physics it seems to be impossible to draw a clear line between the a priori and the a posteriori, and the tangle into which Kant gets himself over this question in the *Opus postumum* only serves to show how difficult the problem is. The problem of the a priori and the a posteriori is too far-reaching to be handled within the scope of this enquiry.

8. H. Dingler, *Das Experiment*, Leipzig 1928; see also Alf. Nyman, *Das Experiment, seine Voraussetzungen und Grenzen*, Z. f. philos. Forschung IV, 1949; C. F. von Weizsäcker, *op. cit.* (§ 8, *n.* 3).

§ 13

1. A. v. Haller, *Die Falschheit menschlicher Tugenden*, 289–290 (Gedichte, ed. Hirzel, Vol. II):

Ins Innre der Natur dringt kein erschaffner Geist,
Zu glücklich, wem sie noch die äussere Schale weist.

2. Goethe, Works I, Weimar 1887, Vol. 3, p. 105:

Natur hat weder Kern noch Schale
Alles ist sie mit einemmale.

§ 14

1. R. Seeliger, *Analogien und Modelle in der Physik*, Studium generale I, 1948, pp. 125*ff*.

§ 16

1. H. Rickert, *Das Eine, die Einheit und die Eins*, Logos II, 1911.
2. Avicenna, *Metaphysics*, ed. Horten, New York 1907, p. 171.
3. Knittermeyer, *op. cit.* (§ 6, *n.* 1).
4. H. Leisegang, *Über die Behandlung des scholastischen Satzes: 'Quodlibet ens est unum, verum, bonum seu perfectum' und seine Bedeutung in Kants Kritik der reinen Vernunft*, Kantstudien XX, 1915, p. 402*f*.—In Philosophische Studien II, Berlin 1950, p. 212, Leisegang declines to recognise this connection, and says it is impossible to build any bridge over the chasm which separates the old ontology from Kant's transcendental philosophy. He refers to the above article, in which he had shown that Kant had the problems of transcendental philosophy in the narrower sense before him in the form which they had taken in Wolff and Baumgarten, and that he had put aside this form of transcendental philosophy as unfruitful. I am in complete agreement with findings of this article. But the question is, whether these arid ontologies are the only form in which Aristotelian and Thomistic transcendental philosophy was still alive. This does not seem to be the case. More recent investigations, notably those of H. Knittermeyer, G. Söhngen, O. Becker, and others, have brought to light new and more fundamental connections, and I was able to confirm these results through my own investigation into William of Ockham. As this question on the one hand involves going deeply into a very large quantity of material, and on the other hand depends on a fundamentally different conception of the history of philosophy, I shall have to reserve my comment.
5. J. Gredt, O.S.B., *Elementa philosophiae aristotelico-thomisticae*, 7th ed., Freiburg in Br. 1937, Suarez, *Met. disp. II*, Mainz 1605, E. Bodewig, *Zahl und Kontinuum in der Philosophie des hl. Thomas*, Divus Thomas 13, 1935.

§ 17

1. E. Husserl, *Ideen zu einer reinen Phänomenologie und phänomenologischen Philosophie*, Halle 1913, § 13. (English translation, *Ideas: General Introduction to Pure Phenomenology*, London 1931.)
2. Duns Scotus, *Reportata Parisiensia*, I,q.24.
3. G. Martin, *W.v. Ockham*, Berlin 1949, p. 44.
4. Ockham, In I Sent., d.30q.3k.

§ 18

1. Leibniz, *Opuscules et fragments inédits*, ed. Couturat, Paris 1905, p. 9.

§ 19

1. See § 20, *n.* 1, viii.
2. See § 20, *n.* 1, xv.
3. See § 6, *n.* 3.
4. See § 20, *n.* 1 xiii*a*.

§ 20

1. The interpretations of Kant discussed in §§ 19, 20 and 21 are as follows:

(i) F. H. Jacobi, *David Hume über den Glauben, oder Idealismus und Realismus*, 1787, in the collected works, Vol. II, 1815.

(ii) C. L. Reinhold, *Briefe über die Kantische Philosophie*, in the periodical *Deutsche Merkur*, 1786–7, in book form 1790–2.

(iii) S. Maimon, *Versuch über die Transzendentalphilosophie*, Berlin 1790.

(iv) G. E. Schulze, *Aenesidemus, oder über die Fundamente der von dem Herrn Prof. Reinhold in Jena gelieferten Elementarphilosophie, nebst einer Verteidigung des Skeptizismus gegen die Anmassungen der Vernunftkritik*, Helmstedt 1792; ed. A. Liebert, Berlin 1911.

(v) J. S. Beck, *Einzig möglicher Standpunkt aus dem die kritische Philosophie beurteilt werden muss*, Riga 1796.

(vi) W. Windelband, *Über die verschiedenen Phasen der kantischen Lehre vom Ding an sich*, Vierteljahresschrift für wiss. Philosophie I, 1877.

(vii) Fr. Paulsen, *I. Kant*, 5th ed. (1st ed., Stuttgart 1898, English translation, by J. E. Creighton and A. Lefèvre).

(viii) H. Cohen, *Kants Theorie der Erfahrung*, 3rd ed., Berlin 1918.

(ix) Br. Bauch, *Kant*, Sammlung Göschen, 2nd ed., Berlin 1916, and *I. Kant*, 2nd ed., Berlin 1921.

(x) E. Adickes, *Kant und das Ding an sich*, Berlin 1924.

(xi) M. Wundt, *Kant als Metaphysiker*, Stuttgart 1924.

(xii) N. Hartmann, *Diesseits von Idealismus und Realismus*, Kantstudien XXIX, 1924.

(xiii) H. Heimsoeth, (*a*) *Metaphysische Motive in der Ausbildung des kritischen Idealismus*, Kantstudien XXIX, 1924, and (*b*) *Persönlichkeitsbewusstsein und Ding an sich in der kantischen Philosophie*, in *I. Kant*, Festschrift publ. by Albertus-Universität, Königsberg 1924.

(xiv) H. Heyse, *Der Begriff der Ganzheit und die kantische Philosophie*, München 1927.

(xv) M. Heidegger, *Kant und das Problem der Metaphysik*, Bonn 1929 (2nd ed., Frankfurt 1951; French translation, Paris 1953).

(xvi) G. Krüger, *Philosophie und Moral in der kantischen Kritik*, Tübingen 1931.

§ 21

1. See § 20, *n.* 1.
2. J. G. K. C. Kiesewetter, *Versuch einer fasslichen Darstellung der . . . neuen Philosophie*, Vol. I, Berlin 1795.
3. See § 6, *n.* 6.
4. See § 7, *n.* 1.

§ 22

1. R. Lehmann, *Kants Lehre vom Ding an sich*, 1878, p. 5.

§ 23

1. Heimsoeth has drawn attention to Kant's usage in this respect in the second of his two essays (§ 20, *n.* 1, xiii), p. 47.
2. E. Adickes, *op. cit*, (§ 20, *n.* 1, x), p. 59.
3. E. Adickes, *op. cit.*, p. 119*n.*

§ 24

1. M. Heidegger, *Sein und Zeit*, 3rd ed., Halle 1931, § 5, p. 18. (English translation, *Existence and Being*, London 1949.)
2. See § 1, *n.* 2.
3. C. A. Crusius, *Entwurf der notwendingen Vernunftwahrheiten*, 2nd ed., Leipzig 1753, § 48.

§ 25

1. Goethe, *Works*, II, Weimar 1887, Vol. 11, p. 47*f.*

§ 26

1. Heimsoeth was the first to bring out the important connections here, in his article *Der Kampf um den Raum in der Metaphysik der Neuzeit*, Philosoph. Anzeiger 1, 1928.
2. *Kants Vorlesungen über die philosophische Religionslehre*, ed. Poelitz, 2nd ed., 1830, p. 53*f.* and pp. 177*ff.*
3. Aristotle, *Met.* XII.6.1071b20, XII.7.1072a25, 1072b8. Aquinas, *com. in met.* 2494*f.*, 2517*f.*, 2524; *s.th.* I,25,1.

§ 27

1. H. Heimsoeth, *op. cit.* (§ 20, *n.* 1, xiiib), p. 44.
2. M. Wundt, *op. cit.* (§ 20, *n.* 1, xi), p. 9*n.*
3. E. Adickes, *op. cit.* (§ 20, *n.* 1, x).

§ 28

1. H. Heimsoeth, *op. cit.* (§ 20, *n.* 1, xiiib), on the knowing subject pp. 6–30, on the acting subject pp. 23*ff.*
2. On the problem of the self, see especially: H. J. de Vleeschauwer, *La Déduction Transcendentale*, Antwerp 1934*f.*, Vol. II, pp. 584*ff.*, and Vol. III, pp. 216*ff.*; Fichte 2. *Einleitung in die Wissenschaftslehre*, chap. 6 (Works, ed. Medicus, Leipzig 1910, Vol. III, pp. 52*ff.*); H. J. Paton, *Kant's Metaphysic of Experience*, London 1936, Vol. II, p. 404, 'The Existence of Self'.

§ 29

1. E. Adickes assembled a great number of such passages, without trying to achieve completeness; *op. cit.* (§ 20, *n.* 1, x).
2. E. Adickes, *op. cit.*, p. 38.
3. N. Hartmann, *op. cit.* (§ 20, *n.* 1, xii), pp. 190*ff.*
4. H. Heimsoeth, *op. cit.* (§ 20, *n.* 1, xiiib), p. 135.
5. H. Heimsoeth has drawn attention to the importance of this reflection, *op. cit.*, p. 131.
6. M. Wundt, *op. cit.* (§ 20, *n.* 1, xi), p. 427.
7. G. Krüger, *op. cit.* (§ 20, *n.* 1, xvi), p. 227.
8. G. Krüger, *op. cit.*, p. 230.
9. G. Krüger, *op. cit.*, p. 234.

§ 31

1. B. Liebrucks, *Platons Entwicklung zur Dialektik*, Frankfurt 1949, p. 20, pp. 79*ff.*, p. 169*f.*
2. P. Natorp, *Platos Ideenlehre*, 2nd ed., Leipzig 1921, chap. XI, especially p. 389.

§ 32

1. F. Paulsen, *Kants Verhältnis zur Metaphysik*, Kantstudien IV, 1900, especially p. 416.
2. H. Heimsoeth, *op. cit.* (§ 20, *n.* 1, xiiia), p. 122*f.*
3. Aristotle, *Met.* VI.1.1028b2: 'And indeed the question which was raised of old and is raised now and always, and is always the subject of doubt, viz. what being is is just the question, what is substance?'

INDEX

www.ingramcontent.com/pod-product-compliance
Ingram Content Group UK Ltd.
Pitfield, Milton Keynes, MK11 3LW, UK
UKHW020212290926
15470UKWH00001B/132

9 780837 171548